Technical and Professional Communication

Integrating Text and Visuals

Technical and Professional Communication

Integrating Text and Visuals

Dolores Lehr
La Salle University

ISBN 13: 978-1-58510-257-0
ISBN 10: 1-58510-257-1

Cover image: © istockphoto/graphixel

Printed in Canada.

10 9 8 7 6 5 4 3 2 1

0209TC

Contents

Foreword

The approach used in this book encourages readers to view both text and visuals as one—an integral part of the overall document rather than as separate entities with the emphasis placed on writing and with graphics viewed as merely add-ons or with the entire layout ignored. To achieve this end, each chapter focuses on some aspect of text and graphics while discussing one or the other more extensively.

While the integration of text and graphics is the main focus of the book, the book also focuses on technical and professional communication as an evolving process that results in products or applications that come about from the interaction between writers and their audience, between writers and peers, and between writers and technologies. Specifically, the book emphasizes the need to first identify one's audience in order to determine the appropriate mix of graphics and text. It also emphasizes looking at every document visually—whether a letter, brochure, proposal, or report—and considering how both text and graphics should appear within it.

The book is divided into **four parts**:

> **Part I** focuses on the planning of documents and includes chapters about aspects that one needs to consider before beginning a document: decisions about its purpose, audience, mix of text and graphics, resources, format, and collaboration. This part also looks at legal and ethical issues, sources for and evaluation of information, drafting text, and sketching graphics or preliminary pages and screens.

> **Part II** focuses on composing text and generating graphics with chapters devoted to writing, designing and laying out pages, creating figures, and using photos and color.

> **Part III** looks at the applications themselves—the integration of text and graphics as they appear in technical descriptions, instructions, proposals, reports, correspondences, promotional materials, and oral presentations.

> **Part IV** serves as a reference for correcting common writing errors, using punctuation, and documenting sources according to MLA and APA guidelines.

Special Features of the book include the following:

- **Integration of Text and Graphics**—the focus throughout the chapters is placed on text and graphics rather than primarily on text. This joint focus is in sync with how most readers approach technical and professional communication today. While studies on visual rhetoric emphasize such integration, unfortunately, most textbooks do not.

- **Emphasis on Identifying Audience**—this book emphasizes the importance of identifying audiences and communicating with people with different levels of technical knowledge. This emphasis appears in discussions within chapters, checklists, and exercises.

- **Business and Marketing Applications**— this book includes chapters not only on producing technical instructions and descriptions but also on producing proposals, reports, correspondence and promotional materials such as guides, flyers, and brochures.

- **Discussions of Electronic Résumés and Posters**—as part of the discussion on correspondence, the book has sections on electronic résumés and cover letters, showing their formats as well as providing current research on what employers are looking for in both. A sidebar in the chapter on oral presentations offers information on presenting posters electronically.

- **Section on Discussion Groups, Blogs and Wikis**—the chapter on correspondence contains a discussion of more recent trends in correspondence and collaboration with such vehicles as discussion groups, blogs, and wikis that allow professionals to share ideas with colleagues, review documentation, and work to create new ones.

- **Applications of Up-to-Date Technology**—also throughout are examples of researching with the Internet, generating tables, charts, and graphics with Microsoft® Word and Excel®, designing pages with Adobe® FrameMaker®, and presenting information with PowerPoint® slides as well as with podcasting, video blogging, and screencasting software.

- **Sample Student Papers**—in addition to showing materials written by professional communicators working for Hewlett Packard, Epson, and other corporations, the book has numerous examples of materials written by students in technical and professional writing classes.

- **Brevity**—this book is shorter in length than other textbooks so that instructors can supplement it with their own examples while still having a full range of topics to choose from in constructing their syllabi. Being under 400 pages, this book is thus both portable and versatile—and offers a solid introduction to both technical and professional communication.

Acknowledgments

Many people have helped to make this book possible. First I'd like to thank my former students who generously allowed me to include their papers from my technical communication and professional writing classes:

- Stacey Kauffman, Kirsten FitzMaurice, and Joe Pelone for letting me use their descriptions, proposals, reports, and accompanying photos and drawings;

- David Sullivan for his outline for a research paper as well as Leanne McMillan and Theresa Lelinski for sections of their feasibility report;

- Donna Mscisz Williams, M.D. for her definition of chemotherapy and her procedure for applying to medical school; Caroline Glenn Maestro, M.D. for her proposal for a study on Attention Deficit Disorder; and Regina Hierholzer for her instructions on using a stemming machine and feasibility report on starting an online floral business.

I'd also like to thank Nick Bernardo for the valuable information he provided on creating proposals, and both Cory Anotado and Jelena Drazenovic who allowed me to photograph their electronic posters.

In addition, I'd like to thank La Salle University for a summer grant and research leave to start this project and my colleagues at La Salle—Jack Seydow, Patricia Haberstroh, and Margot Soven—who recommended this project for funding. Professor Haberstroh, in addition, encouraged me from the beginning to write this book; and Professor Soven advised me throughout the process.

I'm grateful also to Bernetta Doane, Research Librarian at La Salle's Connelly Library, for allowing me to use her CARDS method for evaluation sources.

Besides these students, colleagues, and friends, I'm indebted to others:

- Al Brown, Joe Broderick, Tim Esposito, Steve Lungren and Dana Madonna from the Philadelphia Metro Chapter of the Society for Technical Communication, who gave me samples of work they produced and helped me obtain permission to use them;

- Martina Courant Rife from the Association for Teachers of Technical Writing, who allowed me to reprint her poster submitted to the ATTW's annual conference;

- Professor Judith Scheffler of West Chester University, who reviewed chapters and graciously shared with me her insights from her many years of teaching technical communication;

- Karen Leet, a former colleague at Temple University, who generously read early drafts and e-mailed her suggestions from her home in Kentucky;

- John Lord, a former copyeditor, who volunteered to review chapters in their final format.

I'd also like to thank my publisher Ron Pullins for his helpful advice and patience while completing the book as well as his staff—especially Cindy Zawalich and Linda Diering for their invaluable editorial and production assistance.

Special thanks goes to Hannah Loper, both a friend and former student, who helped with obtaining copyright permissions, proofreading chapters, and revising parts of the manuscript.

Finally I thank my daughters Maria and Maureen, who patiently followed my progress, proofread several versions of this book, and cheerfully supported me throughout its entire production.

PART I:
Planning Documents

This section deals with decisions that one needs to make before creating a document as well as considerations that will affect the document's publication. It contains four chapters that focus on these decisions.

Chapter One

———

Getting Started

Chapter Two

———

Looking at Legal and Ethical Issues

Chapter Three

———

Gathering and Evaluating Information

Chapter Four

———

Drafting and Sketching

CHAPTER ONE

Getting Started

What is meant by technical and professional communication? Definitions of this term can be shown by looking at its purpose, audience, language, and use of graphics. Although most people who write about technology seldom pursue a career in technical or professional communication, they, nevertheless, need to make some preliminary decisions before they actually begin to create any technical documents. These decisions involve determining the purpose, audience, mix of text and graphics, format, resources, and means of production.

Objectives

- ✓ Learn what different forms technical communication may take.
- ✓ Define technical communication by purpose, audience, language, and mix of text and graphics.
- ✓ Understand how determining purpose and audience affects other decisions.
- ✓ Learn about other decisions that need to be made before producing documents.

Defining Technical and Professional Communication

Technical and professional communication is a very broad term. You can describe almost any type of writing connected with technology and business as technical or professional communication. While you might associate it with writing that accompanies industrial and consumer products, much of the writing done for the Internet today is considered technical or professional communication. The following is just a brief listing of what you might classify under it:

- manuals
- instructions for consumers on how to install or assemble a product
- reports to colleagues and supervisors on one's activities or studies
- proposals for projects
- specifications for a product
- scripts for training videos
- handbooks of procedures and policies
- web pages detailing rules and regulations

While this list contains many different examples, you might understand what technical and professional communication is more fully if you look at the purpose of and audience for most technical documents, and its characteristic style with a mix of text and graphics.

Technical and Professional Communication Defined by Purpose and Audience

Technical and professional communication seeks to give information that readers need and sometimes to provide an analysis of it as well. Whether it's defining a term, describing an operation, reporting on a recent trip, or proposing a study, its purpose is to communicate some needed information that often allows readers to take action. Unlike creative or expository writing, technical and professional writing does not seek primarily to entertain, to reveal truths about our existence, or to spark the imagination. While not always didactic, this communication often instructs while it provides necessary factual information.

The audience for technical or professional communication is often more clearly defined than that of creative or expository communication. With different levels of technical content, the communication is usually sent to different individuals or groups of people who seek needed information.

Technical and Professional Communication Defined by Language

The language of technical and professional communication is perhaps one of its most distinguishing features. Unlike that of creative writing, which employs imagery—metaphors, similes, and figures of speech—to convey an imaginative view of reality, technical communication uses factual language to give an unbiased view of an object, process, or concept. Most of this language consists of *simple, everyday words*—mostly nouns and verbs; and *descriptive adjectives*— imagery, metaphors, similes, and other figures of speech—are absent. Also, *numbers* are prevalent. In addition to the diction, the style consists of a very *straightforward syntax.* Most sentences are, in fact, somewhat short, though for a more technically-advanced audience, they can be longer and more complex.

Technical and Professional Communication Defined by the Use of Graphics

Along with language, the most distinguishing feature of technical and professional communication is the integration of graphics or design elements within the text. Unlike creative or expository writing, professional and technical writing uses a variety of graphic devices to communicate fully. These include visuals such tables, charts, diagrams, and other illustrations as well as design elements such as headers, rules, type, bullets, and use of color. Also, the effective use of "white space" within pages is characteristic of this type of communication. (See Chapter Seven for a more detailed explanation of "white space.")

Making Decisions

While you may understand what technical communication is, before you begin to write or produce any of the numerous documents that are classified as part of it, you need to make some preliminary decisions about your specific document—its purpose, audience, mix of graphics and text, and format. You also need to consider what external or internal resources and means of production are available.

Deciding What Your Purpose Is

As explained earlier, your *general purpose* is *to inform or analyze.* You inform about a project, analyze a lab, give instructions on how to complete a task, and explain how to assemble a piece of furniture, or show how to troubleshoot a system to correct programming errors.

Before creating a document, you want to determine not only what the general purpose is, but also a more *specific* one–what *specific goal* you want to accomplish. For example, perhaps your general purpose is to inform while preparing a set of instructions; however, your specific purpose might be to show the reader how to find and delete an unwanted file using the Safe Mode feature of a computer. Or while your general purpose might be to analyze information you gathered for a feasibility study, your specific purpose might be to examine the different costs and features for determining whether or not a fast food restaurant should open on a campus site.

Deciding Who Your Audience Is

In addition to determining your specific purpose, you need to identify the *primary reader— and any additional ones*. For example, you might put together a proposal for a new product that will be read not only by your immediate supervisor but also by people from the marketing, manufacturing, quality control, and legal departments of your organization. Such multiple audiences will focus on different aspects of your proposal based on their individual knowledge and experience. Consequently, your understanding of the audience will lead to several other decisions:

- the level of technicality of your writing
- your choice of graphics

To understand this audience, however, consider first creating a profile of them. (See questions that follow.)

Creating a Profile

Profile of Technical and Professional Readers

When planning a document for an audience, the writer must first consider the background and current position of the reader. Some questions to ask include the following:

Multiple Readers

- Is there only one main reader, or are there several people who will read this document?
- If there is more than one, will the other readers read it as thoroughly as the main reader?
- If there is more than one reader, will the other readers read different parts or will they focus on different aspects? If so, what are these?

Current Position

- What is the current position of the readers of the document?
- Do they hold supervisory roles, or are they peers?
- Do they have the authority to accept or reject all or part of the document?

Background

Personal

- What is the age group of the reader(s)?
- Are they males or females?
- What is their cultural, ethnic, or racial heritage?
- What is their political slant?
- What is their formal education?

Professional

- In what area have they been trained?
- What is their knowledge of the topic of this document?
- What is their experience with this topic?
- What is the likelihood that they will continue to be involved with the topic?

Diction for Different Audiences

You can identify such different levels of technical knowledge for audiences as "high," "medium," and "low." An audience with a high degree of technical knowledge will be comfortable with diction that is more formal and includes jargon that this audience readily understands. On the other hand, an audience with a low degree of technical knowledge will need a simpler, less formal type of diction and will need any jargon explained or substituted with more commonly used terms (see Table 1-1).

Table1-1: Level of Technical Knowledge and Diction

High Level	Low Level
More formal diction	Less formal diction
Technical jargon	Simple everyday words

Deciding Which Graphics to Use

An analysis of your audience will also help with another decision: the selection of graphics. A more technical audience will be comfortable with detailed schematics, exploded drawings of mechanisms, and intricately involved tables and graphs. On the other hand, a less technical audience may need more graphics that are simpler in their design and more readily comprehended, such as pie charts (see Table 1-2):

Table 1-2: Level of Technical Knowledge and Graphics

High Level	Low Level
Complex schematics and drawings	Simple drawings and photos
Intricate tables and multiple graphs	Informal tables and simple graphs

Determining the Mix of Text and Graphics

After identifying the technical level of your audience and selecting the appropriate graphics, consider what graphic and design elements will be used, the number of them, and how they will be integrated within the text.

Before and Now

Decades ago in the mid-eighties, before the Desktop Publishing revolution started and the Internet became readily available, professional and technical documentation consisted of pages with mostly text and a very limited number of illustrations. Professional artists lacked current software programs and needed to spend many hours to create even the simplest line drawing. Photographs, if used, were quite expensive because professional photographers had to be hired, and any changes to photographs involved a very time-consuming and tedious airbrushing. Clip art, actually rudimentary drawings that were "clipped" from pages of books purchased by art departments, was sometimes used. Space around text was not valued as essential, and consequently lines of text were packed together on a page with very little leading between them. Figure 1-1 on the next page provides an example of such documentation produced in the 1980s.

However, with the birth and widespread popularity of desktop publishing, this situation changed. With the new digital equipment came a need to learn how to install, operate, and repair

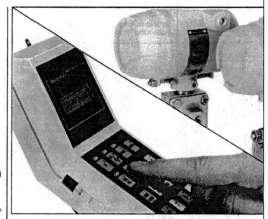

Figure 1-1: Specification sheet created in the 1980s for a process control device. *Reprinted courtesy of Honeywell International, Inc.*

it. Thus, reading instructions and other types of manuals became more important. And with the availability of software that coincided with the rise of desktop publishing, graphics became relatively inexpensive to produce.

As a result, more and more pages of technical documentation included fewer words and eventually more graphics. Also, as companies became increasingly international, it was to their advantage to use graphics that, unlike text, did not need to be translated, and these graphics could be comprehended by people who spoke different languages. Now while most current technical documentation still uses some text, the text is more spread out on a page and mixed with graphic elements.

Following is an example of such a page from a recent technical document that uses a large amount of white space and much less text than documents published two decades ago. As you can see, this document has much more space between and around the text—and so much less text than the earlier one. While both documents have a single graphic, the graphic in the more recent one is larger, and the recent document also has color, bolding of text, and icons.

Setting User Security Permissions

The level of access for each user in the system is set in the User Security module. Permissions can be assigned to allow specified Users to view and modify Diary and Task entries of subordinate personnel, or to allow specified Users to view and modify all Diary and Task entries in the system. Access to User Security settings is restricted to System Administrators and other authorized persons.

In the Utilities menu, open the Security submenu and then click Users to view the User Security window.

Setting User Security Permissions for Tasks

1. In the User Security window, select a User Profile Type in the list at the top of the left pane.

2. In the left pane, select a User name.

3. click [**Permissions**] to view the Categories for assigning Permissions.

4. To assign a Permission to the User, select an item in the Category list and then click [**Enable**]. Repeat this step to assign more Permissions.

5. To save the User Security Permission settings, click 🖫 Save.

Setting User Security Permissions

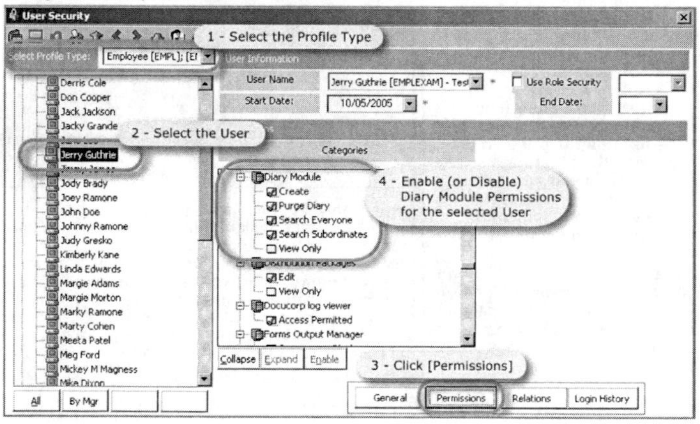

Figure 1-2: Technical document published after 2000. *Reprinted courtesy of IDP (Insurance Data Processing), Inc.*

Subject and Audience as Determining Factors

While generally the number of graphics and graphic elements has increased in documents produced in the last decade, what determines the mix for any particular document are often the subject or content and audience's familiarity with it.

Most likely if your audience is either familiar with the subject of the document or unfamiliar with the language in which your document is written, you will use fewer words and more graphics. However, if your audience is unfamiliar with the subject but familiar with your language, even if it's technical, you'll probably use more words and fewer graphics (see Table 1-3). However, the kind of words and graphics you use will depend on the technical level of your audience.

Table 1-3: Amount of Text and Graphics

More Text and Fewer Graphics	Less Text and More Graphics
Audience familiar with the language	Audience unfamiliar with the language
Audience unfamiliar with the product	Audience familiar with the product

Deciding on Format

Another decision you'll need to make is about what form your document should take. Keeping in mind where the document will be used—and by whom— should help you determine its format. For example, a small spiral-bound manual might be sufficient for an audience with a high degree of technical knowledge who would not need so many details about the product's features. But, if your audience is involved in the purchasing process and not technically knowledgeable, you might put the information in an attractive four-color brochure.

However, not only the level of technical knowledge, but the age and experience of your readers will influence your choice too. For example, an older and experienced reader might be more comfortable with a traditional format such as a manual or brochure with sufficiently large print, whereas a younger reader possibly would prefer a more innovative format and not be concerned with the type size of your document.

Determining Resources

Besides those decisions about the document itself, you'll need to decide about resources—those within your organization and those outside. You'll want to consider materials, equipment, and people. If you're collaborating on a document or having it reviewed, you'll need to consider whom to work with and how to schedule this work.

Internal and External Resources

Within your organization, you can look for information from colleagues and from previously created documents. You might look for information from other writers, engineers, and marketing people who know about the product or a similar product. You might also use equipment or software you're writing about to gain some actual experience with it. Furthermore, to produce your documentation, you need to consider your in-house capabilities and equipment—computers, scanners, cameras, copiers, and binding equipment, for example.

Outside your organization, you might look for information from experts on the subject you're writing about as well as from printed materials and the Internet. For producing your documentation, you might consider hiring a photographer and free-lance writer or editor. You also may want to consider what printing facilities are available and their cost.

Collaborators and Reviewers

At the same time you determine your resources, you'll want to know whether you'll be creating the document alone or collaborating with others. Often in both small and large companies, several people collaborate to write documents. Sometimes this writing may be in the form of specifications and descriptions supplied by engineers or listings of features that marketing people want included and emphasized. Other times, the engineers, software developers, marketing people, and those in quality control might actually take part in composing the document. In any case, you may want to collaborate with others and will need to analyze the collaborative process in a manner similar to your analysis of your audience.

To do so, consider who the individual collaborators are and what expertise each brings. Then, consider who among them is best qualified to write certain parts of the document and how the collaboration will take place.

Who is Involved in the Collaboration?

When you know that you'll be collaborating, it's a good idea to learn something about the people with whom you'll be working. Thus, you may want to ask some of the following questions about your collaborators and their skills:

Background
- Who are the writers and illustrators?
- Whom do they report to?
- What do you know of their writing, graphic, and computer ability?
- What do you know of their training?
- Are they comfortable working with others or just independently?
- What is the protocol for working with them?

Contributors
- What are their areas of expertise?
- What do they individually know about the subject of the document?
- From what perspective will they be looking at the document?

Identifying and Assigning Tasks

After such an analysis, determine what the tasks for the collaborators are. These can be single tasks or multi-layered. In any case, early on in the process the head of the team should assign responsibilities that fit well with each collaborator's background and skills. Also early on, he or she should hold a meeting with the collaborators and give everyone a list of specific responsibilities and tasks.

Scheduling

The head of the team should also arrange these tasks in order and assign dates for their completion. The best way to approach this scheduling is to work backwards:

1. First, list the date the document must be completed in its entirety and set an earlier date to allow for any unexpected delays.

2. Then, arrange the tasks in reverse chronological order, and assign dates to each of them so they become steps that eventually lead to the document's completion.

3. Finally, put the name of the person who is responsible for each task or step next to it.

Deciding about Production

Often within publication departments or small companies, you'll need to decide on how your documents will be produced and distributed. For print and online documentation, obviously your choices will differ.

Decisions for Print Documents

Paper: One decision you need to make is what type of paper you'll print your document on. For correspondence and reports, you can use regular 20 lb. copy paper; however, for promotional pieces and books or booklets, you'll need to consider other types of paper. If you're having the document printed or duplicated externally, printers should provide you with a selection of coated and uncoated paper of different weights. If you're duplicating internally, consider what size, weight, and type of paper works with your copiers.

Folding and Binding: If your document has multiple pages, you'll also have to decide how it will be folded and bound. A flyer, of course, will not be folded; however, a brochure is usually folded in threes with an accordion-type fold or sometimes the front and back panels of the brochure are folded inwards. Short reports are usually held together with staples or clips, but longer ones are often spiral-bound or put into three-ring binders. If there is a large number of pages that need to appear as a book, printers will use "perfect binding," which means they will glue the pages to the cover and spine.

Distribution: Finally, you'll want to consider how you'll distribute your document. The binding and weight of your paper is a concern if you're mailing your publication. Costs for mailing have increased significantly over the last decade, so be wary of using a large-sized, heavyweight paper for an extensive mailing, even if you do qualify for a special rate. If you're distributing the documents in small quantities or are distributing them internally, then the size, weight, and binding of your pages are not as much a concern as they would be otherwise.

Decisions for Online Documents

If your document is an electronic one, you'll need to consider its format and distribution also. Microsoft®, Lotus Notes® and other programs provide templates for e-mail and Web pages from which you can choose. Whatever layout you use, you'll want to keep in mind the same principles of design that you use for print documents, such as repetition, balance, alignment, and proximity. (See a discussion of these principles in Chapter Seven.)

Formatting correspondence and Web pages: While e-mails are similar to memos and have replaced them in the business world as the main means of internal communication, keep in mind that with e-mail you need a salutation and complimentary close (unlike a memo) in addition to the subject line in the body of the message. Web pages consist of links, tables, and headers. They provide a ready means of communicating, but they do require some knowledge of Web authoring languages (HTML or XML) and software programs such as Adobe® GoAlive® or Adobe® Dreamweaver®. However, you can convert Microsoft® Word documents to Web pages by using the "Save as" function.

Distribution of electronic documents: Distributing online documents is much easier than distributing printed ones. There are no addresses to be typed on labels; no wrapping and transporting to a mail carrier. Using *e-mail*, you can readily send your message and document to a large or small group of people whose online addresses you've previously assembled in an address book. However, keep in mind the following:

- Verify that your address book is up-to-date so that your e-mail reaches everyone in the group.
- Consider if you want to copy others as well.
- Keep your message concise and to the point.
- Use attachments for lengthy messages or documents.
- Also, consider using such mail options as adding a signature or requesting a return notice, if your system has these capabilities.

For web pages, be sure to follow the protocol for posting them and the guidelines given by the organization's domain to which you're adding your site.

Checklist

	Do You Know	Yes	No
1.	what the general purpose is and how it distinguishes technical communication from other types of communication?		
2.	how audience helps to define technical communication?		
3.	how the language of technical communication differs from that of other types of communication?		
4.	how technical communication is defined by its use of graphics?		
5.	how the mix of text and graphics has changed over the last two decades?		
6.	what specific purposes you might have for different types of documents?		
7.	how different audiences can affect decisions about documents?		
8.	why the mix of text and graphics can vary with different purposes and audience?		
9.	what decisions you need to make regarding format?		
10.	the difference between external and internal resources?		
11	what to keep in mind when selecting collaborators and reviewers?		
12.	the process for assigning tasks and scheduling?		

	Do You Know	Yes	No
13.	what decisions need to be made for production of print documents?		
14.	what decisions need to be made with electronic documents?		
15.	why the distribution of documents is important to consider?		

Exercises

1. Name several types of documents that you would classify as examples of technical or professional communication. What characteristics do they have in common?

2. Determine the general and specific purpose for each of the following:
   a. a letter of application for employment
   b. a proposal for additional software purchases
   c. an e-mail to other employees assigned to a particular project requesting a team meeting

3. What questions should you ask about your audience before you prepare to create a document? Why is sketching an audience profile helpful?

4. Find an article on the American Heart Association website (http://americanheart.org), and explain with examples what diction is used and how the sentences are constructed. Then go to the Heart Information Network site (http://heartinfo.org) and find another article. Again, explain with examples what diction is used and how the sentences are constructed. How do the two articles differ, and who is the audience for each?

5. In this last site and two others—for example, NASA (http://astrobiology.arc.nasa.gov) and Government Security Act Regulations (http://www.publicdebt.treas.gov/gsr/gsrsecrg.htm)— analyze the mix of text and graphics. Do words or graphics dominate? What types of graphics are used, and are they effective?

6. Working with another student, determine the best format for each of the following (in doing so, explain how you would distribute each):
   a. a notice about a meeting to discuss the company's new health insurance carrier and plan
   b. a floral shop's list of hours and its offerings
   c. an offer of employment
   d. a notice about a change in pay periods
   e. information on a club at school
   f. information designed to recruit new members

7. List as many resources as possible that you would use for a feasibility study about each of the following:
   a. adding a major in your college curriculum
   b. obtaining a room for your student activity on campus

   c. increasing membership in a Greek or other student organization.

8. Set up a team among your peers to collaborate on a project. Who will be on your team? What are their assignments, and how will the collaboration and review cycle work? How might you schedule assignments for your project if it needs to be completed three months from today?

9. Discuss how you would bind each of the following:
   a. 150-page report
   b. training manual of 30 pages
   c. 5-page proposal
   d. 230-page book
   e. letter with a technical specification sheet

10. How might you distribute technical information online? What would be the advantage of distributing it electronically rather than in print?

CHAPTER TWO

Looking at Legal and Ethical Issues

Legal concerns have always been considered important for professional and technical communicators. However, with the scandals involving large corporations like Enron and WorldCom in recent years, more emphasis is being placed on ethical concerns as well. This chapter looks at some of the major legal and ethical issues that you may have when creating documents or using software, the Internet, or graphics.

Objectives

- ✓ Distinguish between legal and ethical concerns.
- ✓ Become aware of general legislation involving copyright protection of print, graphics, and digital materials.
- ✓ Understand what constitutes an ethical dilemma and decision.
- ✓ Appreciate the need for complete accuracy in reporting information verbally and visually.

Legal Versus Ethical

Legal concerns in technical and professional writing are generally those that relate to laws governing the use of materials created by others. These are mostly copyright issues involving infringement as stated in two main acts: The Copyright Act of 1976 and the more recent Digital Millennium Act of 1998. Some professional communicators also have concerns related to the First Amendment.

Ethical issues, however, are those gray areas where behavior might not actually violate the law but, nonetheless, appear objectionable. As Lori Allen and Dan Voss in *Ethics in Technical Communication: Shades of Gray* define the term, ethics is "doing what is right to achieve what is good."[1]

Legal Concerns

The Copyright Act of 1976

According to the U.S. Copyright Office, "Copyright is a form of protection provided by the laws of the United States…to the authors of 'original works of authorship,' including literary,

---

1 p. 83.

dramatic, musical, artistic, and certain other intellectual works." The 1976 Act provides owners with certain rights:

- "To reproduce the work...
- To prepare derivative works...
- To distribute copies or phonorecords...
- To perform the work publicly...
- To display the work publicly...
- In the case of sound recordings-*[sic], to perform the work publicly by means of a digital audio transmission."[2]

What Can be Copyrighted

Any original work can be copyrighted as long as it exists as "a tangible form of expression."[3] Thus, you can copyright a letter, poem, story, graphic sketch, painting, musical score, statue, engraving, or film. However, if you have an idea and have not expressed it in a tangible form, copyright law does not protect it. Neither does the law protect information that is considered common knowledge or "[t]itles, names, short phrases, and slogans: familiar symbols or designs; mere variations of typographic ornamentation, lettering, or coloring; mere listings of ingredients or contents."[4]

How to Copyright

Some people assume that to copyright their works, they need to register them with the U.S. Copyright Office, but that assumption is incorrect. Actually, the very act of having created something tangible gives you copyright protection. In other words, once you create something that physically exists, you've automatically copyrighted it.

Previously, under the 1909 Act, you had to publish your work or register it with the Copyright Office in order to have it copyrighted. With the 1976 law, however, publication was extended to "the distribution of copies or phonorecords of a work to the public by sale or other transfer of ownership, or by rental, lease or lending."[5] Hence, you can "publish" now by distributing a copy of your work to anyone.

Also, while giving notice of a work being copyrighted—a requirement under the previous 1909 act—was still a requirement of the original 1976 act, in 1989 the United States adopted the ruling of the Berne Convention, which protected works without notice. Giving notice of copyright, nonetheless, is still a good idea to alert others who might be unaware of the current copyright law that your work is copyrighted. To give this notice, you need three items:

---

2   "What is Copyright?" Copyright Office Basics. *U.S. Copyright Office*. July 2006. 11 May 2007 <http://www. copyright.gov/circs/circ1.html#fnv>.

3   "What Works Are Protected?" Copyright Office Basics. *U.S. Copyright Office*. July 2006.11 May 2007 <http://www. copyright.gov/circs/circ1.html#fnv>.

4   "What Is Not Protected by Copyright?"Copyright Office Basics. *U.S. Copyright Office*. July 2006. 11 May2007 <http://www. copyright.gov/circs/circ1.html#fnv>.

5   "Publication" Copyright Office Basics. *U.S. Copyright Office*. July 2006. 11 May 2007 <http://www. copyright.gov/ circs/circ1.html#fnv>.

- "symbol © (the letter in a circle) or the word 'copyright'…
- year of first publication of the work…
- name of the owner of the copyright."[6]

This type of notice is for "visually perceptible copies," whereas the symbol "P" in a circle is for a recording. While it's no longer necessary to give notice or register your work with the Copyright Office, you must, nevertheless, send two copies of a work to the Copyright Office at the Library of Congress within three months of your copyrighting it. Although formal registration with the Office is not required, doing so can offer extra protection if some day you take legal action for copyright infringement.

Fair Use

The 1976 Copyright Act includes several limitations to the rights of owners provided by copyright protection. Section 107 contains one such limitation called "Fair Use" that allows the use of copyrighted material without the owner's permission for commentating, reporting, and educational use. This section does not specify what amount of work can be used without permission, but consideration needs to be given to the proportional amount of the entire work that is used—and it should be relatively small. Another consideration is whether using the material will reduce the commercial value of the work and, if so, to what extent.

The Digital Millennium Copyright Act of 1998

Obviously, since 1978 when the provisions of the 1976 Copyright Act took effect, new types of media had gained prominence; and thus new legislation to protect its owners was needed. This new legislation appeared as the Digital Millennium Copyright Act, which Congress passed on October 14, 1998, and President Clinton signed two weeks later. This act, which initially music producers sought in order to keep their works from being copied via the Internet, focuses on several forms of electronic media in addition to music CDs. It protects owners of software, video production, Internet sites, and other digital forms.

According to the U.S. Copyright Office Summary, the act "implements two 1996 World Intellectual Property Organization (WIPO) treaties: the WIPO Copyright Treaty and the WIPO Performance and Phonograms Treaty" as well as "a number of other significant copyright-related issues."[7] These issues include the following:

- "limitations on the liability of online service providers for copyright infringement,"
- copying a computer program for "maintenance or repair,"
- distance education,
- " 'webcasting' of sound recordings on the Internet,"
- "transfers of rights in motion pictures."[8]

---

6   "Form of Notice for Visually Perceptible Copies." Copyright Office Basics. *U.S. Copyright Office.* July 2006. 11 May 2007 <http://www. copyright.gov/circs/circ1.html#fnv>.

7   "The Digital Millennium Act." Copyright Office Summary. *U.S. Copyright Office.* December 1998. 11 May 2007 <http://www. copyright.gov/legislation/dmca.pdf>.

8   "The Digital Millennium Act." Copyright Office Summary, *U.S. Copyright Office.* December 1998. 11 May 2007 <http://www. copyright.gov/legislation/dmca.pdf>.

First Amendment

In addition to copyright laws like those discussed earlier, the first amendment to the United States Constitution can affect your work in technical and professional communication. While this amendment guarantees free speech and the right to publish, it does not allow you to make false statements that can harm another individual or organization.

False statements or statements that can harm another's reputation are called slander—if spoken—and libel, if published. While you probably won't be placed in a situation where libel is likely to occur, if you write news items about people, you need to be aware of the potential for libel suits.

Ethical Concerns

Ethical concerns are those involving behavior that does not directly violate any law but is less than completely honest and fair. A classic case involves the Challenger disaster, where those in charge ignored communications sent about potentially harmful conditions because they wanted to launch the shuttle as originally scheduled, without delays. While in this instance no one was held legally responsible, often unethical behavior, especially if it can harm others, will lead to legal action.

Values as the Basis for Ethical Behavior

If you suspect unethical behavior in your work environment, you face a dilemma and need to make an ethical decision to resolve it. According to Lori Allen and Dan Voss, you can resolve ethical dilemmas by analyzing values—"the underlying values that are colliding to determine the most appropriate course of action."[9] Often, however, you'll be tempted to rationalize a course of action that is unethical because of failure to acknowledge the conflicting values or interests, or because you might exaggerate the consequences that will come from choosing a more ethical course.

Your personal values, as well as those of the organization where you work, will determine the ethics of your communication. Thus, you don't want to work for an organization whose ethical code, whether written or non-written, is in conflict with your own beliefs.

Codes of Behavior

Many professional organizations such as the STC (Society for Technical Communication) and the IABC (International Association of Business Communicators) have guidelines or codes of ethics to which its members subscribe (see Figures 2-1 & 2-2). However, whether or not you belong to one of these groups, you need to have a clear sense of where you stand on ethical issues.

---

9   *Ethics in Technical Communication: Shades of Gray* (New York: John Wiley & Sons, 1997), p. 16.

International Association of Business Communicators

Code of Ethics for Professional Communicators

Preface

Because hundreds of thousands of business communicators worldwide engage in activities that affect the lives of millions of people, and because this power carries with it significant social responsibilities, the International Association of Business Communicators developed the Code of Ethics for Professional Communicators.

The Code is based on three different yet interrelated principles of professional communication that apply throughout the world.

These principles assume that just societies are governed by a profound respect for human rights and the rule of law; that ethics, the criteria for determining what is right and wrong, can be agreed upon by members of an organization; and, that understanding matters of taste requires sensitivity to cultural norms.

These principles are essential:

- Professional communication is legal.

- Professional communication is ethical.

- Professional communication is in good taste.

Recognizing these principles, members of IABC will:

- engage in communication that is not only legal but also ethical and sensitive to cultural values and beliefs;

- engage in truthful, accurate and fair communication that facilitates respect and mutual understanding; and,

- adhere to the following articles of the IABC Code of Ethics for Professional Communicators.

Because conditions in the world are constantly changing, members of IABC will work to improve their individual competence and to increase the body of knowledge in the field with research and education.

Figure 2-1: First page of the Code of Ethics for Professional Communicators. *Reprinted courtesy of the International Association of Business Communicators.*

STC Ethical Guidelines for Technical Communicators

As technical communicators, we observe the following guidelines in our professional activities. Their purpose is to help us maintain ethical practices.

Legality

We observe the laws and regulations governing our professional activities in the workplace. We meet the terms and obligations of contracts that we undertake. We ensure that all terms of our contractual agreements are consistent with the STC Ethical Guidelines.

Honesty

We seek to promote the public good in our activities. To the best of our ability, we provide truthful and accurate communications. We dedicate ourselves to conciseness, clarity, coherence, and creativity, striving to address the needs of those who use our products. We alert our clients and employers when we believe material is ambiguous. Before using another person's work, we obtain permission. In cases where individuals are credited, we attribute authorship only to those who have made an original, substantive contribution. We do not perform work outside our job scope during hours compensated by clients or employers, except with their permission; nor do we use their facilities, equipment, or supplies without their approval. When we advertise our services, we do so truthfully.

Confidentiality

Respecting the confidentiality of our clients, employers, and professional organizations, we disclose business-sensitive information only with their consent or when legally required. We acquire releases from clients and employers before including their business-sensitive information in our portfolios or before using such material for a different client or employer or for demo purposes.

Quality

With the goal of producing high quality work, we negotiate realistic, candid agreements on the schedule, budget, and deliverables with clients and employers in the initial project planning stage. When working on the project, we fulfill our negotiated roles in a timely, responsible manner and meet the stated expectations.

Fairness

We respect cultural variety and other aspects of diversity in our clients, employers, development teams, and audiences. We serve the business interests of our clients and employers, as long as 1, such loyalty does not require us to violate the public good. We avoid conflicts of interest in the fulfillment of our professional responsibilities and activities. If we are aware of a conflict of interest, we disclose it to those concerned and obtain their approval before proceeding.

Figure 2-2: Beginning of the Ethical Guidelines for Technical Communicators. *Reprinted courtesy of the Society for Technical Communication.*

Applying Laws and Ethics to Text

In technical and professional communication, it's important to adhere to the copyright laws and to obtain permission for borrowed materials from their owners. Even if you quote or paraphrase material created by another person under the doctrine of "Fair Use," you need to acknowledge that person or source in a citation. Appendix C provides some examples for citing sources using the Modern Language Association (MLA) or American Psychological Association (APA) style guides. But to find more detailed and specific examples, go online to such sites as Purdue's Online Writing lab (OWL) at http://owl.english.purdue.edu/owl/ or Long Island University B. Davis Schwartz Memorial Library's site at http://www.liu.edu/cwis/cwp/library/workshop/citation.htm.

Unlike copyright infringement, ethical violations occur not by disregarding sources or breaking laws. Rather, they occur when materials are changed or information is deleted so that the text does not reflect the whole truth.

Often with technical writing, the question of ethics comes into play when you have important information that you might include in a document, but if included, might delay the production of the product or might make the product less attractive. An example might be information about possible damage to the product if it's used incorrectly or under different conditions. A more serious situation occurs if omitting important information from a document might harm the user of the product.

Applying Laws and Ethics to Graphics

When using graphics that have been created by others, be careful that you have the permission of the owner or that they are not protected by copyright and thus are in ***the public domain.*** The following explains what is meant by this term:

> "If a book, song, movie or artwork is in the public domain, then it's not protected by intellectual property laws (such as copyright, trademark or patent law)—which means it's free for you to use without permission."

As a general rule, most works enter the public domain because of old age. This includes any work published in the United States before 1923. Another large block of works are in the public domain because they were published before 1964 and copyright was not renewed. (Renewal was a requirement for works published before 1978.) A smaller group of works fell into the public domain because they were published without copyright notice (copyright notice was necessary for works published in the United States before March 1, 1989). Some works are in the public domain because the owner has indicated a desire to give them to the public without copyright protection.[10]

Some graphics on the Internet are available without cost, but many are not. While you can use most of these graphics for a school project because of the "Fair Use" doctrine, you still need to credit your sources. However, even if you credit the sources, you violate their copyright if you attempt to publish them online without the permission of the owners.

Modifying graphics also can be a violation of the copyright law unless you change them to the extent that the originals are no longer recognizable. For example, Figure 2-3 shows two photos of a computer with the second photo having the brand name removed and another name put on. This photo has been changed but not to the extent that it shows a different product.

---

10 "The Public Domain." Copyright and Fair Use. Stanford University Libraries. 2004. 11 May 2007 <http://fairuse.stanford.edu/Copyright_and _Fair_Use_Overview/chapter8/index.html>.

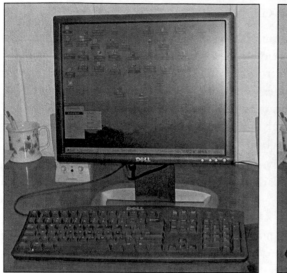

Figure 2-3: Modified photos of a computer monitor and keyboard with the brand name changed in the second one.

Modifying graphics or passing them off as your own is unethical even if they are in the public domain. It's also unethical to create **graphics that misrepresent** what they are intended to show, for example, a chart (Figure 2-4) to represent sales revenue of cell phones with drawings disproportionate in size.

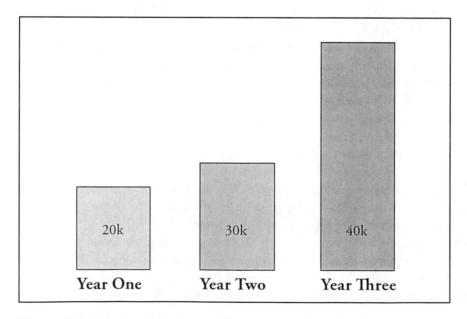

Figure 2-4: Disproportionate graphic representation of sales revenue.

Checklist

	Do You Know	Yes	No
1.	the difference between a legal and an ethical issue?		
2.	what can be copyrighted?		
3.	how to copyright?		
4.	what portion "Fair Use" allows you to use from another's work, and what it does not allow you to use?		
5.	why unethical behavior can lead to legal problems?		
6.	what the basis is for ethical behavior?		
7.	why codes of behavior are important to establish?		
8.	how to recognize an ethical dilemma?		
9.	why omitting information is not only unethical but can be dangerous to others?		
10.	why it's wrong to modify a graphic or misrepresent information graphically?		

Exercises

1. Give several examples of what you can copyright and what you cannot copyright.

2. Whom does the Digital Millennium Copyright Act protect, and why do you think it was needed after the Copyright Act of 1976?

3. How does the First Amendment affect writers?

4. Explain whether the following use of graphics is ethical or legal and why:
   a. Airbrushing the image of a current product for a brochure, to resemble a new product that is not yet ready for release, but is expected to be ready by the time the brochure is printed and distributed.
   b. Changing a table with the sales figures for the past year so that stockholders and board members are not disappointed in the year's revenues.
   c. Representing increases in the number of employee benefits per year for the last five years in a bar graph by using one point increments, so that the number of benefits currently available would seem to have been increased significantly.

     d.   Using disproportionately-sized images (e.g., moneybags) in a pictogram in place of actual amounts to indicate the costs to employees of their yearly benefits.

     e.   In a line chart showing revenue that will appear in a company's annual report, omitting the points on the lines where revenue was negative.

5. With other classmates, discuss the legality and ethics of each of the following situations:

     a.   You take supplies home from your office.

     b.   You submit expenses for reimbursement to your organization for a conference that you attended with a friend, which includes the entire cost, even though the friend paid part of it.

     c.   You apply for employment where you're asked for your salary at your last place of employment. You know that this new position pays more, and you want to get an increase. Since you know that your previous salary can't be verified, do you put down the actual amount or a larger one? What values are in conflict here in this ethical dilemma?

     d.   You're disappointed that an instructor has given you a lower final grade than you expected in a course, and she explains that the project you submitted the last week of class was poor and lowered your average. You explain that when you submitted that project, you sent an e-mail asking her to let you know if there was any problem with the project so that you could revise it; and she acknowledges that she received your message but never answered it. Was your instructor acting ethically and legally in lowering your grade? Why or why not?

     e.   From past experience, you believe that your supervisor is biased against you and looking to find evidence of your incompetence. While working late at night on a project that he gave you insufficient time to complete, you accidentally jam the copier. Do you tell him and your coworkers the next day that you broke the machine? Do you deny you broke it if asked about it? Do you do nothing, and leave your supervisor and colleagues to wonder who broke the copier? What is the dilemma, and how would you resolve it ethically?

6. Explain the meaning of each of the following:

     a.   fair use

     b.   intellectual property

     c.   public domain

     d.   libel

     e.   slander

     f.   copyright infringement

7. Research a situation in the news wherein some public figure is accused of an ethical violation. Then write one or two paragraphs summarizing the situation and explaining why the person seems to have violated ethics rather than the law.

8. City and local governing groups often propose legislation whereby their representatives can't hold positions with companies that have city or state contracts. If you agree with such legislation, write a paragraph supporting it and pointing out possible ethical dilemmas that might occur or future violations of law if it's not passed. If you disagree with this legislation, write a paragraph explaining why such employment is neither unlawful nor likely to be unethical.

CHAPTER THREE

Gathering and Evaluating Information

This chapter looks at how you gather and evaluate information for your technical documents. It also looks at sources of information such as public records, correspondence, interviews, surveys, and site visits.

Objectives

- ✓ Understand the use of both the library and Internet for getting information.
- ✓ Learn to evaluate the reliability of both text and graphics.
- ✓ Learn the use of public records, correspondence, and site visits for gathering information.
- ✓ Understand the process of interviewing.
- ✓ Learn the types of questions used in interviews and surveys.

Researching Information

The Library Versus the Internet

You might think that with the popularity of the Internet and such search engines as Google and Yahoo that you no longer need to use the library. Or if you use it, you need to use it electronically via the Internet. Admittedly, most libraries—especially college and university ones—do have online capabilities that allow its users to search for books and periodicals or read articles without actually having to enter the library building itself. Nonetheless, while this situation indeed does exist, you'll find numerous volumes of materials available only in print. However, since you'll most likely need current information for technical communication projects that you find in recent periodicals, this chapter will look at accessing periodicals mainly online.

Primary Versus Secondary

Primary sources are those original sources that serve as the basis for critical interpretation and which provide you with information to form your own judgments and analyses. These sources consist of letters, e-mail, and other types of correspondence as well as public records (e.g., birth and death certificates or deeds), and lists of data. They also include interviews, surveys, experiments, and site visits. *Secondary sources*, on the other hand, are materials that are based on primary sources and are usually interpretations of these sources. Examples of these would be articles and books. Traditionally, you would find these sources in the library, but now you can also find many secondary sources online. While you may not think of primary sources as library documents, nevertheless, you can find some primary sources, such as letters, in libraries. While researching information you might use some secondary sources, but most of your information will be from primary sources—and probably gathered online.

Finding and Evaluating Information Online

Using Search Engines

Every year more and more Web sites appear with information that you can use in research. To access these sites you can use popular search engines like Google or Yahoo. Google, in fact, has maintained its prominence among all of the others and now offers Google Scholar, another search engine that allows you to find articles published in journals and presented at conferences.

While these popular search engines can be helpful, keep in mind that they are supported by advertisers whose advertisements compete for prominence in a listing. Consequently, a site that can provide the information you need might not appear as prominent as you would expect it to be.

Using Online Subscription Services

Most libraries subscribe to search engines that are accessible only to their members. These include First Search, Lexis Nexus, and ProQuest, which provide abstracts and, in some instances, full text or illustrations for articles. Other bibliographies that previously existed only in print are now available online, such as the *MLA International Bibliography, The New York Times,* and *The Wall Street Journal*—just to name three. Access to these fee-based sites allows you to find books and periodicals that are unavailable through commercial search engines.

Evaluating Internet Sites

Evaluating sources means determining the validity of a site and whether the information on it is accurate. Sites found on university subscription services are more likely to be reliable because they consist of articles and books that have been "peer-reviewed," that is, they have been examined by colleagues who understand the substance of the work and vouch for its validity.

However, most sites found on the Internet through search engines such as Google have not been peer-reviewed. While many are valid, some are not; and your responsibility when researching information is to determine the validity of the site to ensure the accuracy of the information on it. Bernetta Doane, Research Librarian at the Connelly Library of La Salle University in Philadelphia, warns students that because there are no restrictions on who publishes on the Internet, you should proceed with caution when incorporating these resources into your research papers. She emphasizes that the Internet is full of all kinds of information, both good and bad. To gauge the validity of a site, consider Doane's CARDS or five-point approach for evaluating sources on Internet sites (Figure 3-1).

Evaluating Graphics

Along with the printed information you can find on the Internet, you can also find graphics— charts, tables, maps, and illustrations. In evaluating them, you can follow the same process as that listed above, but you might also ask additional questions such as those given in Figure 3-2.

Sources of Information

Public Records

One type of primary source is public records. This source includes census documents, transfer of properties, births, deaths, and sale transactions. These records can provide you with information about a person, organization, or site.

Correspondence

Furthermore, you can often find helpful information in letters or even e-mails, especially those that come with replies to your inquiries. Other correspondence can provide unexpected information about your topic.

CARDS: Criteria for Evaluating Resources via the Internet

Bernetta Doane, MLS
Connelly Library, La Salle University

Credibility—Is there an author(s) or producer(s)? Does the document provide credentials on the author? For example, the author has a Ph.D. or a M.D. Is there contact information for the authors(s) or producer(s) of the document? This can include an e-mail address and/or street address and phone number.

Accuracy—Does the site appear to be comprehensive? Is the information included in the site complete and accurate? How does the information found in the site compare to other resources or related sites?

Relevancy—Is there a stated purpose or information describing the function of the site?

Dates—What is the date of coverage for the site? When was it created? Is the site up-to-date? Are the links up-to-date?

Sources—Does the site offer or link you to additional sites or resources? Are the links appropriate for the research topic? Are these resources easily accessible?

Figure 3-1: Questions to ask when evaluating sources on Internet sites.

- Is the graphic original; and if not, is the source of the graphic given?
- Does the graphic seem to represent accurately what it is intended to represent?
- Is the information given complete?
- Do the colors used provide sufficient contrast, and increase readability?
- Are the numbers and/or callouts fully legible? (See Chapter Five for a definition of callouts)

Figure 3-2: Questions to ask when evaluating infographics on the Internet.

Interviews

Conducting interviews is another profitable way of getting information in technical communication. Doing so allows you get the answers to specific questions and often gain access to unsolicited information.

In-Person, Telephone, and E-Mail

An **in-person** interview is the most common and preferred type of interview because it allows face-to-face contact and can provide additional information to what you originally sought. Also, when people converse, they convey information by facial expressions and gestures that are not readily apparent in other types of interviews. Most people, if they have the time, prefer in-person interviews.

Telephone interviews are used often and can be quite effective, especially if the person you're interviewing has a limited amount of time to spend with you. Although this type of interview lacks the added dimension of facial expressions and body language, nevertheless, the speaker's voice inflections, pauses, and modulations in tone can provide additional clues to understanding what he or she is saying.

Interviewing via **e-mail** is also a means of gaining information, especially from someone who is inaccessible because of distance or time constraints. While e-mails lack the added dimensions of voice and body language found with telephone and in-person interviews, they can be quite effective in that they deliver the requested information in a short amount of time and provide written answers that you can review and directly import into your documents.

The Process

What follows is an overview of the process for interviews, where *preparing* applies to all types of interviews, while *conducting* applies only to the in-person interview.

1. Preparing

Determining purpose and researching interviewee: Before attempting to interview someone, you want to determine what you hope to accomplish. Also be sure you're interviewing the appropriate person who can answer your questions. Try to research this person's background so that you're better prepared to formulate appropriate questions and to respond to answers from him or her.

Contacting: Next, contact this person and make an appointment. In doing so, be sure to give the person an estimate of how long you'll be meeting—whether it's 15 to 20 minutes or 30 to 40 minutes. Most people are very busy; and if someone agrees to let you interview him or her, you don't want to take more of this person's time than is absolutely necessary. Make sure also that you ascertain the exact time and location of the interview and, if necessary, directions to the location.

Formulating questions: After you've made the appointment, you want to draft a series of questions before arriving. While it's fine to have some *close-ended questions* (see "Surveys— Types of Questions"), most of your questions should be <u>open-ended</u> to allow the person to talk freely about the topic. In fact, you might use short close-ended questions as a jumping off point to allow the person being interviewed to elaborate on or explain why he or she gave a specific answer.

When you've devised your questions, consider putting them in order. To do so, consider a logical progression—possibly going from those requiring the least difficult answers to those requiring the most difficult ones, from the general to the specific, or from one topic to another.

Writing questions: After you've formulated your questions, consider writing them on note cards (one question per card), or listing them clearly on a page of your note pad. Whatever you do, have them handy and ready to be read with a glance so that your questioning will not seem too mechanical or impersonal.

Reviewing: Finally, as you prepare to conduct your interview, review the questions again and possibly read them aloud so that when you're ready to use them, they will be familiar to you and sound somewhat spontaneous. If you're conducting a personal interview, decide what you'll bring—a notepad or recorder—how you'll dress, and how you'll arrive at the designated location.

2. Conducting a Personal Interview

Be sure to arrive on time for the interview and preferably a few minutes early. Begin by thanking the person for his or her willingness to speak with you, and then proceed to your questions. If you wish to record the conversation, ask permission from the person before beginning. If the person objects, be gracious and prepare to take notes instead. If the person doesn't mind being recorded, place the tape recorder in a less-than-obvious spot so that its presence will not inhibit the person.

Begin with easy or simple questions: Begin with questions that the person can readily answer—easy, direct, and non-controversial ones. Your immediate goal is to put the person sufficiently at ease so that he or she will tell you what you want to know, so pose questions first that the person feels comfortable answering. For example, you might ask about some award or achievement that the person received or a project that you know the person supports. After the person responds readily to these questions, then move on to more difficult or more complex questions.

Listen rather than talk: While you, of course, need to keep the conversation moving in the direction you want, try to speak as little as possible, and instead listen attentively to what the person being interviewed is saying. If you find that the person seems taciturn or reluctant to expand on the subjects you broach, you might try interjecting a few brief comments to provide some feedback, or even tell something briefly about your own experience, to encourage the person to speak more extensively. However, you never want to talk more than the person being interviewed about the subject being asked. Encourage the person to expand on a topic if you think he or she can—and don't be afraid to add questions not on your list or note cards that seem to follow naturally from answers. Likewise, don't be afraid to eliminate questions that don't seem to work into the conversation or that might have been answered earlier.

Conclude with verification and appreciation: Be aware of the time; however, avoid looking directly at a clock or your watch. When you see that there are five minutes or less remaining in the time agreed upon, begin to wrap up the interview. If you've not covered all your questions in the allotted time and want the person to answer a specific one before you leave, then move quickly to that one. Tell the person that you appreciate his or her making time to talk with you and that you still have questions but will ask just one more. If the person is willing to extend the interview, then you can ask other questions; but if not, then proceed to your final question.

End by asking if you might verify the accuracy of anything that is unclear in your notes or on the tape (if you recorded the interview), and then thank the person for his or her time.

Reflecting and Responding: After you return from the interview, try to find some quiet place where you can look over your notes or tape and write out clearly the answers to your questions. If there are parts that are unclear, mark them so that you can either clarify them yourself or return to the interviewee for clarification. Try to write out as much as you can recall from the interview—and do so soon afterwards. Psychologists say that we can successfully chunk information from our short-term memory to be passed into our long term one, if we do so shortly after we initially processed this information.

Surveys

Purpose and Benefits of Surveying with Questionnaires

Like an interview, a survey is a fine method of gaining primary information. However, unlike an interview, it usually consists of much shorter questions in the form of a questionnaire. Because the answers in questionnaires are often brief and standardized, unlike the answers to interview questions, you can quantify them more easily and put the results in tables or charts.

Types of Questions

Like questions for interviews, those in questionnaires for surveys can be either **open-ended** questions, which solicit a variety of answers for any one question, or **close-ended**, which solicit only one answer from among several choices. Some typical close-ended questions are as follows:

Multiple choice: These are questions with several (at least three) answers from which the person completing the questionnaire must choose one.

Either/Or: These questions are often choices between one alternative and another: for example *yes* or *no*.

Ranking: Such questions ask the respondent to put a list of items in their order of importance or some other logical order (e.g., *least likely, somewhat likely, more likely, most likely*).

Length

Most questionnaires should be rather short if you want a respondent to complete them. Preferably, they should be just one page; but depending on the complexity of the subject and the likelihood that respondents will fully complete them, they can be longer—even extending to several pages.

Online vs. Printed

The length of your questionnaire can be a factor in determining whether you'll distribute it online or in person. If it's long, and you decide to distribute it online, consider how much scrolling is required. Of course, you can avoid unnecessary scrolling with "skip logic" questions that will direct the respondent to another screen and another part of the questionnaire.

Online questions are easy to distribute with e-mail and, in many cases, require less work to complete. They are most effective if you need to reach a large number of respondents whom you would not be able to contact personally within a short time. Some online services, like surveymonkey.com, offer the opportunity to post surveys free if the number of responses needed is within their guidelines (for example, 100 responses). However, for a very reasonable monthly fee, services like surveymonkey.com allow you to send out your survey to a larger number of people (for example, 1,000) and to create as many surveys as you like. For a sample on this service's demo survey, see Figure 3-3.

Design and Layout

The design and layout of your questionnaire is very important and will vary depending on whether you have an online or printed one. Consider the following:

- **amount of space around text.** You want the respondent to have sufficient room to answer questions, particularly if you include open-ended ones requiring short paragraphs (see "White Space" in Chapter Seven).

- **alignment.** Make sure items in multiple-choice answers are aligned directly with similar items in other questions.

- **similar grammatical structures** in the phrasing of all your multiple-choice answers. In other words, be sure that they consist of the same grammatical constructs —nouns or noun phrases, verbs or verb phrases, or clauses (see Chapter Five for an explanation of parallel structure).

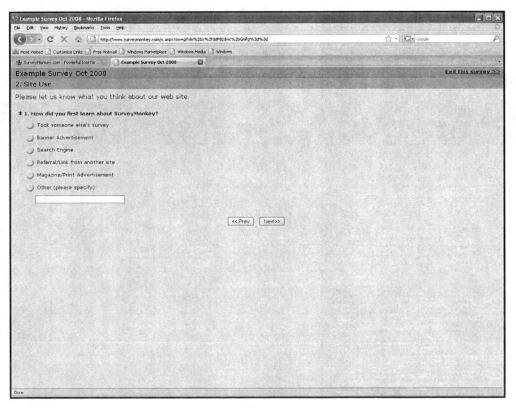

Figure 3-3: Sample question in surveymonkey.com's demo. *Reprinted courtesy of surveymonkey.com.*

- **use of an appropriate type font.** You want a font that is large enough for your readers to see and one that is appropriate for their age and culture.

The main thing to keep in mind is that you want your questionnaire to be attractive to the reader so that he or she will take the time to complete it.

Site Visits

Determining Purpose

Site visits are a fine way of gathering information that is not readily available from other sources. It allows you to experience first hand the actual situation that you're studying. It also allows you to gain an additional perspective on a topic and sometimes to learn about aspects that you previously didn't consider.

Before visiting a site, consider what you hope to accomplish. Ask yourself what you'll be looking for specifically and how you'll obtain that information. While you might find some unexpected information when you visit, you want to have a clear idea of what you hope to learn before seeing the site.

Recording Observations

Be prepared when you visit a site to record your responses, whether by using an electronic device, taking written notes, or checking items on a form. If you were visiting a parking lot to see how many spots were being used by employees of a company, you might come prepared with a blank form that you can fill in and complete there. If you're visiting a site just to get an overview of it, then consider recording your reflections using a VHS or DVD camcorder as you look upon it.

Checklist

	Do You Know	Yes	No
1.	why you might need to use a library for research?		
2.	what constitutes primary versus secondary sources?		
3.	what are some major search engines?		
4.	the advantages of using subscription services over popular search engines?		
5.	what to look for when evaluating an Internet site?		
6.	what the acronym CARDS means?		
7.	what to look for in evaluating graphics online?		

	Do You Know	Yes	No
8.	several different sources of information?		
9.	what you might learn from using public records?		
10.	why all interviews are not necessarily conducted in person?		
11	the steps in preparing and conducting an interview?		
12.	the value of reflecting or responding after conducting an interview?		
13.	the difference between open- and close-ended questions and the benefits of each?		
14.	different types of close-ended questions?		
15.	when a site visit would provide helpful information?		

Exercises

1. Give examples of what you would use the library for and what you would use the Internet for in your research.

2. List several sources that you consider primary and several you consider secondary. Can you list any that might be both, if used differently?

3. What are the advantages of using fee-based databases rather than free ones?

4. Locate *two* web sites that contain information that might help you with a project for this course or another one. Then evaluate each by answering the following questions about the source and the evidence or information it provides.
   a. Is it up-to-date, reputable, and trustworthy? Why or why not?
   b. Is the information sufficient, balanced and reasonable?
   c. Can the information be verified?

5. If there are graphics on these sites, evaluate them by answering the questions given earlier in Figure 3-2.

6. Construct a list of questions for a brief (5-minute) interview of another student. Interview the student, using your questions. Did you need additional questions? Did the student answer the questions as you expected him or her to answer them? Discuss.

7. What are the advantages and disadvantages of each of the following:
   a. open-ended and close-ended questions in a questionnaire?
   b. printed and online questionnaires?
   c. site visits?

8. Name three points to keep in mind when designing and laying out a questionnaire.

9. In conducting a site visit, what might you do beforehand and while at the site?

10. Identify what would be the appropriate means of gathering information for each of the following, and explain why in one or two paragraphs:

    a. Determining the feasibility of building another school parking lot.
    b. Deciding whether or not to build a convenience store at a specific location near your campus.
    c. Investigating whether to have a fast food restaurant on campus.
    d. Researching whether to add courses to an existing minor discipline or to create a new minor.

CHAPTER FOUR

Drafting and Sketching

As part of planning, you need to make preliminary drafts of your text and sketches of your graphics. To get started, you might try some popular prewriting techniques—brainstorming, free writing, mapping, outlining, and storyboarding. This chapter discusses each of these techniques and how you might begin to sketch visuals and design pages before actually starting to compose.

Objectives

- ✓ Learn different methods of prewriting to prepare a preliminary draft.
- ✓ Learn what aspects you need to consider when laying out pages and screens.
- ✓ Understand the importance of sketching graphics before beginning a document.
- ✓ Learn how you can prepare a preliminary draft with both text and graphics.

Generating Text By Prewriting

As discussed in Chapter One, you need to consider your purpose and your audience before you begin writing. Then you want to consider what you want to say and how you want to say it. While you might think that you can just begin writing after some reflection, you're wiser if you take some time to actually first "prewrite" about what you want to say. The following are various forms of prewriting that you can use to get started.

Brainstorming

Brainstorming is one of the most common ways to generate ideas. The process is a simple one. You simply jot down ideas on any given topic or, if you prefer, make lists of them. By writing anything that comes to mind on a sheet of paper, you capture your thoughts on a particular topic. Later, you can arrange them into some preliminary structure to use for your document. See Figure 4-1 below for an example of brainstorming.

Installing a door		
Measuring	Closer	Storm windows
Attaching sweep	Adjustments	Phillips-head screws and
Screw covers	Handle components	screwdriver
Door latch	Screen	Window latches

Figure 4-1: Brainstorming for instructions on installing a door.

Free Writing and Quick Writing

In the 1960s, Peter Elbow popularized the form of drafting called "free writing." In its purest form it consists of writing on a topic without stopping—or pausing only to reread and pick up an idea to continue, and then loop back again with that idea. The advantage of this method is that you begin to actually write on your topic and thus begin to formulate what you want to say. The disadvantage is that often when you've already expressed your ideas in sentences and paragraphs, you're reluctant to go back and write about them again. Nonetheless, using this method can sometimes get you to think more critically about the topic and begin to organize ideas coherently.

Mapping

Mapping, unlike free writing, works well if you prefer visuals to text because it allows you to see on paper the relationships among different ideas and different parts of a document. To begin mapping, try the following process:

1. Draw a block or circle in the center of a page, and write a word or phrase in it.
2. Draw other blocks or circles, and put words associated with the word in the central block or circle within these additional shapes.
3. Then, using lines, connect these blocks or circles.
4. Continue drawing blocks or circles, and put words inside them that suggest parts of your previous group of words.
5. Connect these to the second group.

The process of mapping continues until all subdivisions and associated parts are shown on the page. Having this map showing the relationships of ideas allows you to start your document with some concept of its structure.

Figure 4-2, for instance, shows the beginning of a map on installing a door.

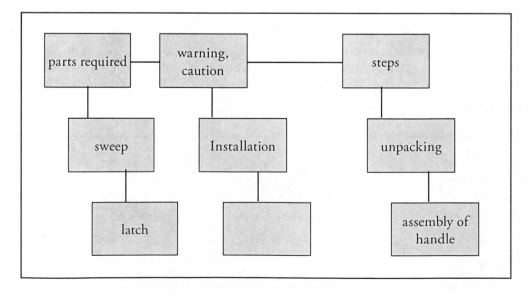

Figure 4-2: Mapping for instructions on installing a door.

Outlining

Outlining is probably the most traditional method of generating text and also one of the most popular visual methods. By indenting words, phrases, or sentences and beginning each with a letter or number, you provide an overview of the relationships among the various parts of your subject. While outlines for books and other large documents can consist of paragraphs, most of your outlines will consist of sentences or phrases.

Create Two or More Sub-Parts

Keep in mind that when you outline, you divide every topic into at least two sub-parts. Thus, if you begin with Roman numerals (I, II, or III), you can't have an "A" without a "B" under one of them. Under the "A" and "B," you can't have a "1" without a "2." (Figure 4-3 shows one possible arrangement.)

Figure 4-3: Structure for an outline with each divided part having at least two sub-parts.

Maintain the Same Structure

Also, keep in mind that all sub-parts need to be *the same grammatical parts*—for instance, all nouns, noun phrases, verbs, verb phrases, and so on. By creating parts of your outline with the same grammatical structure, you're creating "parallel structure" (see Chapter Five for more explanation of this term). Having parallel structure not only makes your outline look attractive, but it also helps when you begin to flesh out your outline when writing your first draft.

Storyboarding

Storyboarding is perhaps the most appropriate method for integrating the visual and verbal parts of a document. By dividing a page into two parts—one for text and one for visuals—or by using two facing pages, you can begin to design the pages for your document. Beginning with your overall idea as your "story," you then put what you conceive to be the first point or chapter of your story on one page or one side, and then list or actually draw the table, chart, or illustration that you think should accompany it (see Figure 4-4). While you do not need to accompany all your points with graphics, you can begin to determine whether you need graphics at all or what graphic elements you might include such as bullets for lists, color-coded rules, and type fonts for different level headers.

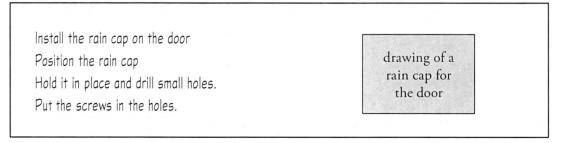

Figure 4-4: Storyboarding for part of the instructions on installing a door.

Laying Out Pages or Screens

While drafting your text, you want to begin sketching your layout. Certainly if you use storyboarding, you can work on both almost simultaneously. However, if you use some of the other methods, you want to begin sketching soon afterward to see what your pages will look like. To do so, consider the following:

- **Headers** - write the name of each header, and leave spaces beneath each. (See Figure 4-5.)

- **Illustrations and Links** - Create blocks, and write the names of any charts, tables or other figures you want to use within them. If you're sketching Web pages, write the URLs for the links to other pages.

- **Style Sheet** - Devise a list of specifications (font name, type size, and paragraph spacing before and after) for each header and sub-header.

- **Space Around Text and Graphics** - Consider how much space you want to leave around sections of text and headers (for a fuller explanation of what is termed "white space," see Chapter Seven).

- **Margins** - Consider how wide or narrow you want your page margins to be.

These are some of the considerations you want to address before you actually begin creating the document.

Unpacking the door
        Pulling out the windows
        Pulling out the sweep
        Pulling out the screen
            Checking hardware
Assembling the door handle

Illustration
of door

Figure 4-5: Sketch of a layout for a page.

Sketching Charts, Tables, and Other Illustrations

Chapter Six focuses in detail on creating effective charts, tables, screens, and illustrations for your professional and technical documents. However, at this initial stage you want to identify those that you might need. So in each of the blocks that you created in your rough layout of the document, go back and provide more details.

- If you're planning to have a bar chart, for example, you might draw one by hand or create one using Microsoft® Excel® or a drawing program such as Adobe® Illustrator®. Likewise, if you need a table, you can create one in Microsoft®Word or at least identify it by name. At this early stage, if you know only the type of chart or table and what you want it to convey, **type its name and any other information you currently have in the box.**

- Likewise, if you need a specific illustration but do not know where to get it or what it will look like, **draw a rough sketch, or type something about this drawing**—its name and purpose—and perhaps add a stock illustration or piece of clip art as a placeholder for the draft. Figure 4-6 is an example of a rough sketch of flowers that a student drew at this preliminary stage of planning a set of instructions on arranging flowers.

- If you need a screen, draw a rectangular block and sketch its contents.

- Also, you want to sketch any other graphic elements you might use: **boxes, rules, arrows, or logos,** for example. Perhaps you want a rule to separate your

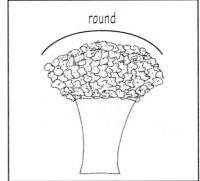

Figure 4-6: Hand-drawn sketch for instructions on how to arrange flowers.

major sections or supply a logo on the top of each page. If so, then go through your pages, and indicate or sketch where each graphic element should appear.

By either sketching or identifying the graphic by name and purpose, you begin to formulate what you want to communicate visually and want to develop later.

Preparing Preliminary Drafts

Now with your pages laid out and your visuals sketched, you're ready to begin your first draft. To do so, once again review your purpose and think of your audience. Then try the following procedure:

1. If you've a map outline or storyboard, on a sheet of paper, list all the major topics or sections leaving plenty of space between each of them. If you've brainstormed or written your ideas in paragraphs, try to extract the principal sections and list them with plenty of space between each.

2. Write as much as you can for each. If you need to stop and gather more information, write a note to yourself indicating that you'll add more information later.

3. As you write, add your sketches of graphics and graphic elements beside the text. (Again, if you've created a storyboard, you should have completed this part of the process.)

4. When you've completed this preliminary draft, print a copy of it to review.

5. As you reread it, take note of sections that are awkward or inaccurate and need more work. Particularly, try to look at the layout and see if your design could be more effective with some changes. Look at your sketches and see how they relate to the accompanying text and the document as a whole.

6. When you've finished your first review, if possible leave the document alone for at least twenty-four hours before going back to draft again.

7.   This time, focus more on the accuracy of what you've written and sketched as well as your style, looking to see if your facts are correct and if your writing could be clearer, more concise, or more coherent in places. Focus also on completing your visuals as well as integrating them within your pages or screens.

When you're finished reviewing this second version, make all needed corrections, and you'll have completed your draft.

Checklist

	Do You Know	Yes	No
1.	why it's important to prewrite to generate text?		
2.	several means of prewriting?		
3.	the process of mapping?		
4.	two important points to consider when outlining?		
5.	the benefits of storyboarding?		
6.	what to consider when sketching pages or screens?		
7.	why preparing a style sheet earlier on is important?		
8.	why sketching graphics or indicating the need for one is helpful?		
9.	what you should look for when reviewing the first draft?		
10.	how long you should wait between preliminary drafts?		

Exercises

1. Select one of the popular means of prewriting (brainstorming, free writing, or mapping), and write for ten minutes on the topic of your next assignment.

2. Keeping in mind the points emphasized about outlining, evaluate the following outline on the reasons for Apple's success with its iPod:

      I.   Introduction
     II. Technology's role in the iPod
         A. Electronics
         B. Decreased size of components
         C. Increased capacity of memory
         D. Unique control mechanism
             1. 4 button, 1 wheel control

   III. Timing

      A. Jobs introduced it in 2001; at the same time, music industry had started to sue those illegally downloading MP3 files.

      B. Also introduced iTunes and the Apple Music store, where consumers could purchase songs to load onto their iPod for just 99 cents.

   IV. Fashionable

      A. iPod is useful, and good-looking

      B. Sleek, simple, white design

      C. Lightweight

      D. White ear buds have become a symbol of status across the world.

   V. Marketing Strategy, the iPod was introduced with powerful U2 television spots.

      1. Simple, 3 Color video spots featured the band performing while a dark figure danced around the screen with the white ear buds and an iPod on his/her hip.

      2. Within 3 months, sales would hit 4 million.

   VI. Conclusion

      A. Jobs' significant impact on the computing world, and emphasize the iPod as his most important contribution to the digital music society in which we live.

      B. End with creative quote and /or phrase.

3. Create a storyboard for a set of instructions on a simple process (burning a CD, starting a car, or something else that you know well).

4. Identify a chart or drawing you'll need for a future project, and provide a rough sketch of it.

5. Explain how you might find topics or section headers for a document if you choose to free write or brainstorm rather than outline, map, or create a storyboard. What might you do if your prewriting consists of a list of either topics or paragraphs with topics within them?

6. Why should you always check the accuracy of your information when reviewing what you've written even within the first preliminary draft?

7. What should you look for when reviewing the second preliminary draft?

8. Prepare preliminary drafts of your next assignment following the steps given in this chapter. Submit this draft, or the second version of it, for a peer review by another classmate.

PART II:
Composing Text and Generating Graphics

This section consists of three chapters that focus primarily on the writing of the document and the generation of graphics. It includes details for writing text, creating figures or tables, and using color and graphic elements.

Chapter Five

———

Writing

Chapter Six

———

Using Tables, Figures and Color

Chapter Seven

———

Designing Pages or Screens and Using Graphic Elements

CHAPTER FIVE

Writing

This chapter focuses on what comprises a technical and professional style of writing and looks at word choice, sentence structure, the use of active or passive voice, and tone. It explains why clarity, brevity, and coherence are important for effective communication in all forms of writing—web pages, manuals, letters, e-mail, brochures, and other print documents. It also looks at writing text to accompany graphics.

Objectives

- ✓ Understand the components that make up style.
- ✓ Appreciate different levels of formality in selecting words for different audiences.
- ✓ Understand the need to use concrete words and action verbs.
- ✓ Learn about different types of sentences and sentence patterns to understand the effect of syntax on writing.
- ✓ Recognize the importance of using parallel structure.
- ✓ Understand when to use active or passive voice.
- ✓ Appreciate the need for clarity, brevity, and coherence.

Style

You can define the term "style" when used for writing as the selection of diction and syntax. Style, however, also includes a writer's voice and tone. As explained in Chapter One, the use of technical words and direct syntax distinguishes style in professional and technical communication. The tone is usually factual and direct, and the voice is mainly active.

Diction

Diction is word choice, and the words that you use in technical communication are both technical and what is termed "plain English." However, while your diction may vary in its level of formality depending on your audience, overall it will be more formal than that of other types of communication.

Levels of Formality

Levels of formality vary from the very formal to the very informal. If you're writing for a very educated, sophisticated audience, your diction will be somewhat formal. On the other hand, if your audience is either less educated or very familiar with your topic, your diction might be more informal.

What distinguishes the level are the words you use—whether they are technical ones with Latin roots or more everyday words with Anglo-Saxon origins. Table 5-1 has several examples of words that basically have the same meaning but differ in their formality.

Table 5-1: Examples of Informal and Formal Words

INFORMAL	FORMAL
Begin	Commence
Grasp	Apprehend
Hidden	Recondite
Think	Cogitate
Break up	Fractionate

Concrete Versus Abstract Words

In addition to classifying diction according to levels of formality, you can classify diction according to whether the words are mainly concrete or abstract. **Concrete** means that the words are names of tangible entities that you can experience through the senses: for example, *chair, desk, book, machine, computer,* or *telephone.* In contrast, **abstract** words are intangible: for example, *faith, organization, concept, transference,* or *intelligence.* While effective diction often contains a mix of concrete and abstract words, you'll use your more concrete words in professional and technical communication; however, the predominance of either type of diction will vary depending on the topic you're writing about and your audience.

Strong Action Versus Weak Invisible Verbs

To be an effective writer, you want to use **strong action verbs** that enliven your sentences and, when possible, avoid using **weak** or "**invisible**" verbs. Examples of action verbs include *write, collect, install, operate, tie, turn, pull, twist, wrap,* and *join.* Examples of weak or invisible include *am, is, are, was, were, has* or *have been.* Compare the following two sentences:

- **Weak/Invisible:** The socket *has* three holes that *are* missing screws.

- **Action:** Three socket holes *need* missing screws.

The first sentence uses two weak verbs, whereas the second has one stronger verb that allows you to be more direct and clear.

Syntax

Syntax is **the arrangement of words in sentences.** While you can arrange words in many different and elaborate ways, you want to use a more direct and straightforward arrangement in professional and technical communication than you would in other kinds of communication. A look at the different types of sentences and sentence patterns can help you understand more fully how syntax affects your writing.

Types of Sentences

Words that appear together constitute **phrases,** and when phrases consist of words that contain a subject and a predicate they become **clauses. Sentences** consist of clauses and vary according to the type of clauses that comprise them.

If a clause expresses a complete thought, it's said to be "independent" or a "main" clause. If it has a subject and predicate but does not express a complete thought and you introduce it with a subordinating conjunction (for example, *after, since, because, if, under, over, when, where,* or *although*), it's said to be "dependent" or "subordinate."

Simple: A simple sentence consists of one independent or main clause. For example: *The installation took several hours to complete.*

Compound: A compound sentence consists of two or more independent or main clauses connected by a coordinating conjunction (*and, but, or, nor, for, so, yet*). For example: *The installation took several hours to complete, and the operation of the system could not start until noon.*

Complex: A complex sentence consists of at least one independent or main clause and at least one dependent or subordinate clause. For example: *Because the installation took several hours to complete, the operation of the system could not start until noon.*

Compound-Complex: A compound-complex sentence consists of at least two independent clauses and at least one dependent clause. For example: *Because the installation took several hours to complete, the operation of the system could not start until noon, and employees were forced to wait.*

While you might use mainly simple or compound sentences in your writing to describe or instruct, if you use complex or compound-complex, you show what action or idea you want to focus on or make more prominent.

Sentence Patterns

Sentence patterns vary from simple, with the subject-verb-object arrangement, to more elaborate and complicated arrangements, such as the Latinate periodic arrangement with the verb coming last. For example: *For all the work that you and your team completed in all the different locations and for all your contributions, which you made so generously, you are to be commended.* As stated earlier, most likely you'll use a straightforward subject-verb-object arrangement in technical communication.

Parallel Structure

To be an effective communicator you also want to use parallel structure in your sentences. Parallel structure means maintaining consistency in the use of grammatical constructs. For example, if you have a list consisting of names of items (nouns or noun phrases), you want to include only names of items—not actions (verbs or verb phrases). Following are examples of a sentence without parallel structure followed by one with parallel structure:

Non-parallel: The technician **completed** the assembly of the new system, **installed** its software, and a **demonstration** of its operation. (*2 verbs and 1 noun*)

Parallel: The technician **assembled** the new system, **installed** its software, and **demonstrated** its operation. (*all verbs*)

Having parallel structure helps your sentences to flow and makes it easier for your readers to comprehend the sentence's meaning.

Voice

Voice has to do with the relation of the subject to the verb—whether the subject performs the action of the verb or is "acted upon"—in other words, whether the action is performed by the object in the sentence or by a doer who is not in the sentence.

Active Versus Passive

When a sentence is in the active voice, the subject performs the action. An example would be the sentence *The technician upgraded the software*. Here the subject *technician* actually performs the action *upgrading*. On the other hand, if the sentence reads *The software was upgraded by the technician* or *The software was upgraded*, the subject *software* is not performing the action. Rather, *technician*, the object in the prepositional phrase, is performing the action or, as in the case of the second sentence, the object *technician* does not appear in the sentence.

Frequently Asked Questions about Voice

Which is preferable? As you can see, active voice is shorter, more direct, and thus preferable to passive voice. Consequently, you want to use active voice as much as possible when it's appropriate.

Should passive voice ever be used? Although active voice is preferable, there are situations where using it would be inappropriate. For example, if the person who performs the action is unimportant or irrelevant to the meaning of the sentence, you would not want to include this person as a subject just to create active voice. For example, it would seem strange if you converted a sentence such as *The rods are inserted after they are heated* to active voice by including a subject: for example, *Someone inserted the rods after he or she heated them*.

Not only do you want to avoid using active voice if the subject—the doer of the action—is unimportant, but also if the doer is unknown. An example would be the sentence *The bill was sent in early March*. In this instance, the doer of the action is not only unimportant to the meaning of the sentence but also is unknown.

Can both active and passive voice be used in the same paragraph? Yes, but the most important point to keep in mind is that while you can use both voices in a paragraph, you want to avoid shifting abruptly from one to another just as you want to avoid shifting abruptly from one voice to another within a sentence. An example of such a shift follows:

> **Wrong:** The technician assembled the system early in the day, and the new software was installed later by him also.

> **Right:** The technician assembled the system early in the day and also later installed the new software.

> The system was assembled earlier in the day, and the software was installed afterwards.

Shifting from one voice to another makes reading more difficult and can confuse your reader about what you're saying.

Tone

Tone is a rather difficult concept to explain in relation to writing except by describing different "tones." When you think of tone, you might think of the sound of a stereo or another audio system. When you talk about the tone of writing, you're actually thinking of how it sounds. Does it sound sad, comic, ironic, or indifferent? The sounds you hear when you read words actually comes from the choice of the words and their arrangement in the sentence. Thus, your tone results from the diction and syntax that make up your particular style.

Characteristics of Style

Clarity

As a professional or technical writer, it's essential that your writing be clear. Numerous examples have been documented where confusion—and in some cases legal action— resulted from unclear writing. To maintain clarity, consider the following:

Use Plain English

While people may differ in what they mean by "plain English," most would agree that it involves using language that is simple and readily understood by most readers. Using plain English would mean you write about "connections," not "interfaces," and also means you write about putting items in order of importance, not "prioritizing" them. Although in technical and professional writing it's sometimes impossible to use simple terms, you want to try to use them whenever possible.

Avoid the Overuse of Jargon

Along with the use of simple terms, you want to avoid using jargon excessively. Jargon is the use of words that are generally understood by a particular group of people. Certainly, as a technical professional, you need to use jargon; otherwise your audience will feel as if you're speaking to them as an outsider. However, since everyone does not understand jargon to the same degree, you want to use it only when necessary and only when its use is likely to be understood by the majority of your readers. Whenever possible, use words that are recognizable by your largest audience.

Define Terms

To help maintain clarity in your writing, you want to define any term that your audience would be unfamiliar with. As explained later in Chapter Eight, you can put a short definition in parentheses after the term the first time you use it, or put the definition in a complete sentence.

Avoid Vague Pronouns

To keep your writing clear, try to eliminate the use of *this, which,* and *it* when they serve to represent a larger concept or idea rather than a specific noun in a previous clause or sentence. An example would be the following: *Students were discussing the ways they could work in groups to complete the project. This was what most of them wanted to do.* Here the pronoun *This* does not go back to any one word in the previous sentence but rather goes back to the idea of working in groups. To correct this error, and make the sentence clearer, you could replace *This* with the phrase *Working in groups.*

This error is called a "broad pronoun reference" because the pronouns should take the place of just one noun, not of a broad idea expressed in a phrase or clause. Like a cowboy who flings a lasso around a steer and pulls him in, when you use broad pronoun references you're throwing a lasso or rope around a whole group of words with an idea buried within them—and then trying to pull that idea into the next clause as one pronoun—"this," "which," or "it."

Use Articles

Although some languages don't include articles, the indefinite articles (*a, an*) and the definite article (*the*) are an important part of English. However, some writers think that they should eliminate them in technical writing. By clipping off these extra words, they believe that they are making their writing seem more "technical." However, in actuality, they are making their communication unclear. Avoid writing sentences such as *put screw in upper-shelf of table,* and write instead *put the screw in the upper-shelf of the table.*

Brevity

Along with clarity, brevity is important in effective professional and technical writing, although you never want to achieve it by eliminating articles or necessary words.

Omit Unnecessary Words and Phrases

Avoid adding unnecessary words to qualify what you're saying. Some examples would include words like *seems, somewhat, really,* and *quite.* Also, in an attempt to clarify your meaning, you may be tempted to add words or phrases to sentences that add little and can be condensed or eliminated: for example the phrase, *In today's ever-changing world.*

Avoid Strung-out Verb Phrases

Avoid especially adding extra words to predicates to create strung-out phrases. These usually consist of parts of the verb *to be* added to the present participial form of a verb. A present tense strung-out verb phrase such as *beginning to study* can be shortened to *begins to study,* just as the present perfect tense passive voice phrase *have been watching* can be shortened to *have watched.* Note that eliminating strung-out verb phrases not only shortens your sentences but reveals a clearer, more definite action.

Avoid Needless Repetition

Also, because of carelessness or because you're desperately trying to be clear, you may add words or phrases that repeat ideas. Examples of these would be *repeat again, type up, finally end,* and *period of time.* Instead delete the unnecessary word, and write just *repeat, type, end,* and either *period* or *time.*

Omit Fillers

Eliminate fillers that begin sentences like *It is* and *There is/are* and replace them with strong verbs. Often you'll find these verbs buried in nominalizations that follow the fillers as shown below:

> <u>There is</u> <u>communication</u> between the doctors and patients about rights to privacy.

> <u>It is</u> a growing <u>realization</u> among the players that they will need to compromise to end the lockout.

If you remove the fillers and uncover the verbs, your sentences are clearer and concise:

> The doctors <u>communicate</u> with their patients about rights to privacy.

> The players increasingly <u>realize</u> that they will need to compromise to end the lockout.

Coherence

The final characteristic of an effective style of writing is coherence. When your writing has coherence, all the parts go together and are readily understandable. To achieve a coherent style, you need to arrange ideas in a logical manner and provide connectives between them so that they flow from one to the other.

Structure Ideas Logically

Some writers think that to achieve coherence you need only supply transitional phrases between ideas in sentences and paragraphs. However, while transitions serve as the "glue" between ideas, if the underlying structure is faulty, they will not hold together no matter how much glue you apply. Therefore, you need to arrange your ideas in such a way that they move in some logical progression. For instance, they can move chronologically from beginning to end, *spatially* from least important to most important, or *emphatically*, by degree of emphasis, with the most emphatic part at the end and the second most emphatic part at the beginning.

Supply Headers

Furthermore, in technical and professional writing you can supply headers to serve as guides to your structure and help readers move from one idea or topic to another. When creating headers and sub-headers, be sure to maintain parallel structure.

Use Transitions

Transitions are the bridges that allow you to move from one idea to another throughout sentences and paragraphs. While you might think of them only as transitional words and phrases, you can create transitions also through the repetition of key ideas, words, and pronouns.

Transitional words and phrases: Transitional words and phrases are those that identify for the reader the kind of relationship that exists between ideas. For example, if you want to show the addition of a new idea to the idea that preceded it, you might use *and, also,* or *furthermore.* If you want to show contrast between two parts, you can use *but, however, in contrast,* or *on the other hand.* To indicate a causal relationship, use *because, since, consequently,* or *therefore.* Other transitional words can indicate an ending (*finally* or *in conclusion*), sequence (*first, next,* or *then*), or comparison (*similarly* or *likewise*).

Repetition: You can also create transitions by repeating words that represent the same ideas or by substituting pronouns for the same person or concept. For example, using words such as *left, leftist,* and *liberal* or the words *right, right-wing,* and *conservative* within a paragraph helps the sentences to flow. So does the replacement of names with pronouns that stand for an idea or person. To repeat a name over and over again makes your sentences choppy, as the following example shows:

> <u>John</u> opened the document, and <u>John</u> marked the <u>pages</u> that were needed; <u>John</u> then counted the <u>pages</u>.

However, you can replace the name with a pronoun and achieve more coherence:

> John opened the document, and he marked the pages that were needed; he then counted them.

Writing Text for Graphics

Headers

Keep in mind that headers serve as graphic elements on a page. While they work effectively to provide transitions between topics, they also serve to break up large bodies of text and allow for space between these topics. In creating headers, consider using a typefont that contrasts in size and type family with the one you use for body text. For example, if you use a 12-point serif font (Cambria, Garamond, or Times Roman) for body text, consider using a 14-point sans serif font (Calibri, Ariel, or Franklin Gothic) for headers. If there is more than one level of headers, begin with a larger point size, possibly 16-point, and then use increasing smaller sizes for the sub-headers.

Captions and Callouts

Captions are summary statements that describe the contents of photos and figures, while callouts are the names that identify parts of objects in photos and figures. When writing **captions**, follow these guidelines:

- Be consistent—use the same type font and punctuation for all descriptions of figures or photos in a document.
- For figures, number them consecutively within a document.
- If you use verbs or verbals, use present tense and active voice.
- If identifying individual parts within a photo or figure, list them from left to right.
- Place the caption either underneath the photo, on the side, or at the top consistently (place titles of tables at the top only).

Callouts are words on photos or drawings that identify particular parts of objects within them. When using callouts, consider the following:

- Use parallel structure (all nouns, verbs, or phrases).
- Place the text on a horizontal line (don't have text "radiating" out from the center on diagonal lines).
- Keep the number of callouts to a minimum so as not to obscure the image. If more are needed, enlarge one part, and separate it for the additional callouts; or use numbers along with a key that matches these numbers to the callouts.
- Use a type font that differs from that of the document's body text.
- Place callouts as close as possible to the parts being identified, and use arrows to point to any parts that are far from their identifying callouts.

For more information on callouts, see Chapter Nine.

Checklist

	Do You Know	Yes	No
1.	what style consists of?		
2.	why concrete words and strong verbs are important?		
3.	what is meant by different levels of diction?		
4.	what type of syntax distinguishes technical writing?		
5.	the different types of sentences and sentence patterns?		
6.	what is meant by parallel structure, and why it's important?		
7.	what voice means, and why active is usually preferred?		
8.	when you can use passive voice?		
9.	why shifts in voice should be avoided?		
10.	what "tone" is in writing?		
11	five steps to create clarity in writing?		
12.	four steps for keeping your writing concise?		
13.	several ways to create coherence in writing?		
14.	the different types of transitions?		
15.	in creating headers, what you need to consider?		

Exercises

1. Make up a list of words and their synonyms that connote more or less formality (e.g., begin—commence; got—received, acquired; eat—partake).

2. List five concrete and five abstract nouns.

3. List five action verbs.

4. Correct the faulty parallelism in the following sentences:
   a. Notify all users, ensure the files are backed up, and the system is then locked.
   b. Power off the monitor and peripherals and the switch of the power supply is put in the off position.
   c. The following installation procedure applies to all phone casing unit models and also, if the phone casing unit is being stacked on top of a phone casing base or another phone casing unit.
   d. Attach the monitor, peripherals, and the casing unit should be attached too.
   e. Position the switch to the on position and power up the monitor; also the system must then be unlocked.

5. Change the following sentences from all passive to all active or, if already active, from active to passive.
   a. If the phone casing base is to be installed in a DRU rack, it is to be installed as per instructions supplied with the option kit.
   b. Remove the operational manual and the foam insert.
   c. The plastic bag protecting the phone casing switch should be removed.
   d. The phone casing switch must be inspected for any visible damage and if any is observed, your distributor/technical support person should be contacted immediately.

6. Identify and correct vague pronouns and missing articles in the following sentences:
   a. Ensure that all air vents are free of dust particles, and that they are one inch (2.54 cm) from side vent and 5 inches (12.7 cm) from rear vent.
   b. When all LEDs on the front panel of the DRU unit are flashing red, this indicates that temperature inside it has reached a point where a failure can occur.
   c. At this point it must be powered down, and this done, ensure that all vents are free of debris.
   d. All LEDs are lit green, which tells operator that system has been turned on.
   e. Check clearance of front vent, and power down system.

7. Rewrite the following sentences by eliminating unnecessary words, repetition of ideas, fillers, and strung-out verb phrases as well as by adding any needed words.
   a. It is recommended that the unpacking procedure be performed prior to starting the installation procedure.
   b. Position the screw that extends from the rear of the lower unit's electronic casing into the stackable phone unit's rear hole that can be found in its electronic casing.

  c. A terminal is one of the methods a user can access the DRU for the lever operation.

  d. There is a screw that is extending from the rear of the unit's electronic casing that is needed to be inserted into the casing.

  e. The on switch on the left of the unit needs to be turned upwards to power on the system.

8. Create more coherence in the following paragraph by supplying transitions:

> When a new signal has just been established, the sender has to start sending at some rate. The sender has no knowledge of the extent of traffic at that time on the network. The rate is known as the first transmittal rate. The rate should be selected so that it is not so high that it causes problems to the network. The rate should not be so low that an application fails with a rate less than it needed, a rate gradually increasing to the preferred rate when the preferred rate would have been acceptable to the network from the start.

9. Write a caption for the Figure 5-1 below:

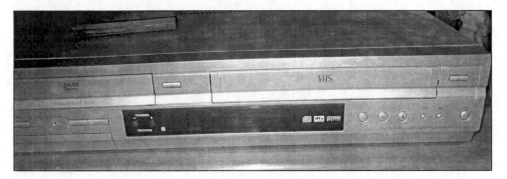

Figure 5-1: _____

       _____

       _____

10. Supply several callouts for the parts of the player shown in this figure, and indicate how they would appear.

CHAPTER SIX

Using Tables, Figures and Color

Illustrations are a very important part of technical and professional communication. To make information readily accessible, you need to integrate graphics within your text to provide a visual blueprint of what you want to convey. This chapter focuses on tables, charts, and drawings—graphics that you're likely to find in technical communication—as well as photos, which appear less frequently. This chapter also looks at colors that are appropriate for graphics and text—for print or online—as well as the use of color for highlighting information and orienting readers.

Objectives

- ✓ Create attractive, accurate tables and learn their uses.
- ✓ Learn about the different types of charts and determine which are best suited for different types of information.
- ✓ Understand the use of other types of illustrations—drawings, diagrams, and maps.
- ✓ Understand the advantages and disadvantages of using photos in technical and professional documents.
- ✓ Learn about the uses of color for print and the Web.

Tables

Uses

Tables are a wonderful means of conveying large amounts of information within a relatively small space. If exact numbers—even those with several decimal places—are needed, then tables are the best choice for a graphic. However, keep in mind that if you have numbers with several decimal places, you want to round them off to two or three places to avoid having a long string of decimals, which can be confusing. Make sure also that you use the same units throughout, and that you indicate what the units are—either with symbols (for example, $) or with a legend or note below the table. Some popular uses of tables are for financial records, salaries, insurance, and taxes (for example, personal income tax tables).

Types

Tables usually contain numeric information, but they can also contain **pictures** or **words**. In technical and professional documentation, they are more commonly used to provide information when you have many numbers and need to use exact numbers. Tables consist of specific parts, which you can add or remove, and are easy to create using any word-processing program.

Parts

Tables consist of **columns** that run vertically and of **rows** that run horizontally. Tables also have **titles**, which always need to be placed at the top, unlike captions for figures, which can be placed at the top, bottom, or side. Headers at the top of columns are called **stub-heads**, and any **notes** related to information in the table should appear in a smaller size font at the bottom, under the table itself.

Creating Tables

Tables are easy to create if you have Microsoft® Word or another word-processing package or if you use a page layout program such as Adobe® FrameMaker®. Using either program, you can create tables with as many rows and columns as you want and format them in a variety of styles ranging from a classic look with grid lines to a more contemporary style, one with colored shading and three-dimensional effects. Table 6-1, for example, was created in Microsoft® Word and formatted as a grid with color accenting and separating the information within it.

Table 6-1: Majors of Students in Prof. Jones' Spring Classes

Majors	English 150	English 108	English 409
Business	3	3	0
Communication	2	3	3
Digital Arts	0	0	3
Education	2	4	0
English	0	0	6
Science	0	1	2
Others	3	2	0

Figures

All illustrations other than tables are labeled as "figures" in the captions that accompany them. The most common types of figures in technical writing are charts, drawings, and diagrams. However, figures can also include photos and screens.

Charts

Uses

Charts are generally used to show comparisons of trends and changes as well as distribution of parts. Unlike tables, charts rarely give a visual representation of exact numbers with decimals, but rather show comparisons with whole numbers of items that change over time, or as percentages of a whole. Three popular types of charts are bar, line, and pie.

Types

Bar Chart: Bar charts show comparisons of amounts. These amounts are usually rounded off to whole numbers so that the solid, often colored, bars convey immediately what they represent (see Figure 6-1). When different colored bars are put next to or on top of one another, the chart becomes a **stacked bar chart** (see Figure 6-2). Like single bars, you can arrange these multiple bars horizontally or vertically. A variation of the single bar chart is a **pictogram**, where you replace the bars with small pictures or icons.

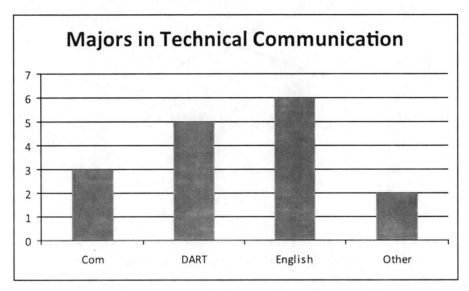

Figure 6-1: Bar chart showing majors in a technical communication course.

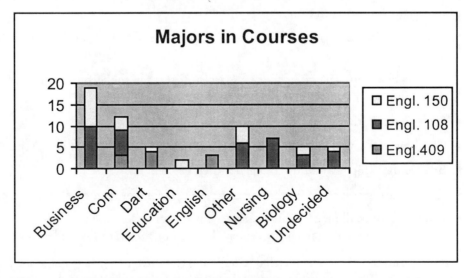

Figure 6-2: Stacked bar chart showing majors in three courses.

Line Chart: Line charts also serve to provide comparisons—especially of changes over time. Unlike bar charts, they can show more exact changes and the immediate impact of change fluctuating up and down. Because of this fluctuation, these charts are sometimes called **fever charts**. Line charts can also show changes in multiple items. To differentiate each item, you can use either different colored lines or solid and perforated ones (see Figure 6-3).

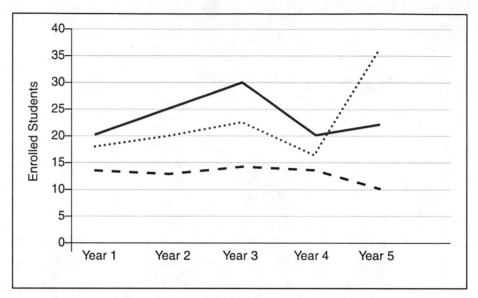

Figure 6-3: Multiple line chart showing changes in the enrollment of three courses over five years.

Pie Chart: Pie charts show the relations of parts to the whole, with the whole being represented as 100%. This type of chart is very effective for showing allotments of a budget, expenditures, geographic distributions, and similar relationships. To create an effective pie chart, keep the following in mind:

- When creating segments, think of the pie as the face of a clock, and begin at the top or 12 o'clock.

- Moving clockwise, create the largest segment first, then the second largest, and then the third, and so on.

- Have no more than 6 or 7 segments.

- Combine very small amounts into one segment, and title it "Other" or "Miscellaneous."

- Be sure that all parts add up to one hundred percent.

- If you're using colors, be sure to use contrasting colors for segments next to each other.

- If you're using callouts, be sure to place your text on a horizontal line—and not a diagonal one radiating from the center. (People don't want to read sideways.)

Figures 6-4 and 6-5 show two pie charts depicting the three major fields of students in a technical writing class. The first one does not follow the guidelines and starts with *Communication,* which is the third largest piece. The second one, however, follows the guidelines and starts with *English,* the largest piece.

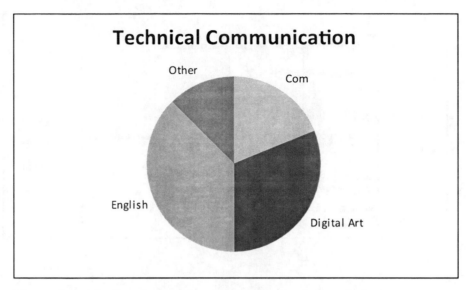

Figure 6-4: Pie chart incorrectly showing majors in a technical communication course with "Communication," the third largest segment, first.

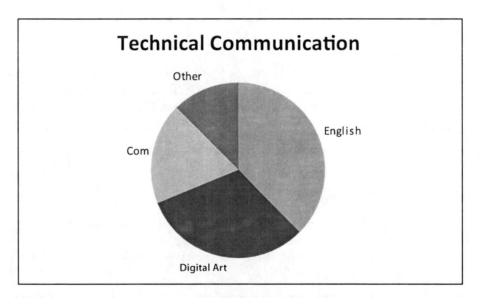

Figure 6-5: Pie chart correctly showing majors in a technical communication course with "English," the largest segment, first.

Creating Charts

Charts are easy to create in drawing programs such as Adobe® Illustrator®. However, you can also create them from your data in Microsoft® Excel® and insert them in your Microsoft® Word documents. Figure 6-6 shows a bar chart created in Excel® and pasted into a Word document.

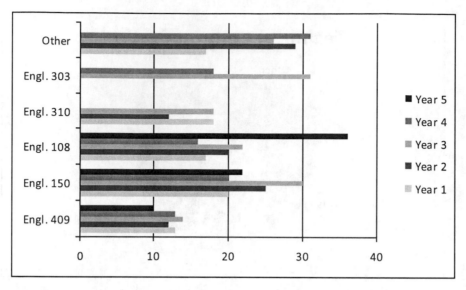

Figure 6-6: Bar Chart created in Microsoft® Excel® from data in a worksheet. *Reprinted with permission from Microsoft Corporation.*

Drawings and Diagrams

Uses

You can use **drawings** to illustrate installations, operations, assembly, and other processes. You also can use them to show parts and various devices. On the other hand, you'll use **diagrams** to explain processes and sometimes relationships, such as those shown in a company's organizational diagram. One popular type is the flow diagram that shows how one part or operation leads to another.

Figure 6-7 is a diagram of the process a student goes through to register for his or her classes.

Creating Drawings and Diagrams

You can create drawings and diagrams in both word processing and drawing programs and import them into your word-processing files. While Microsoft® Word provides a variety of geometric shapes and arrows, its most recent version also offers "SmartArt," in its Illustrations ribbon on the Insert tab (see Figure 6-8).

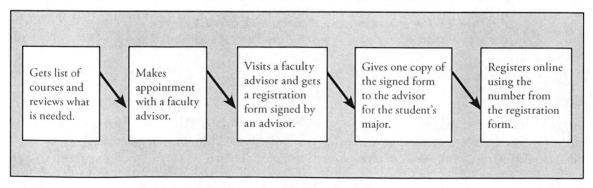

Figure 6-7: Flow diagram of a process for registering for courses.

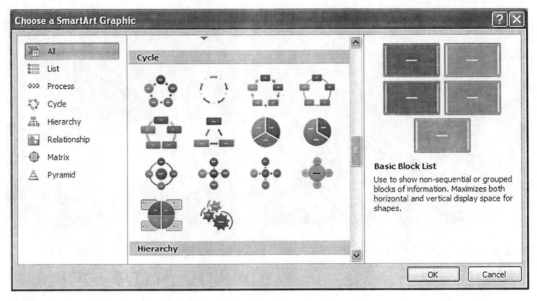

Figure 6-8: Illustrations in the SmartArt option in Microsoft® Word. *Reprinted with permission from Microsoft Corporation.*

Likewise, in FrameMaker® and other page layout programs you can find standard shapes such as ovals or rectangles. To draw more elaborate illustrations, you should use a drawing package such as Illustrator® or Adobe® Freehand®.

Photos

While drawings and diagrams show the details that you often need when describing an object or explaining a concept, you can use photos to give a sense of verisimilitude to the object being depicted. Consequently, you'll most likely use photos prominently in brochures, data sheets, and other promotional pieces to show what the object or product actually looks like. While photos serve this function, they usually have too many extraneous details within them to be used for explaining operations, installations, or other processes. Photos in Figure 6-9 are typical of those that would appear in a brochure where you would want to give a prospective customer a sense of what your products look like, rather than explain their set-up or operation.

You can purchase stock photos, or even download some royalty-free ones. If you have your own photos, you can import them into your computer using photo imaging software like Adobe® Photoshop®. When bringing them into your documents, however, consider their resolution and their size since large files might affect the printing of your document. If you use a program such as FrameMaker®, you can import photos in such a way that they become part of the original file and, thus, reduce the document's overall size.

Screens

Screens, like photos, provide a sense of verisimilitude and show what the reader would actually see, and like photos are used in brochures, data sheets, and other promotional pieces. You can readily capture images of your computer screens by using the "print screen" function found in most word

Figure 6-9: Photos of a screen and printer that would be appropriate for a brochure.

processing programs and then inserting them into documents. To use screens in instructions, you might, however, use a program such as TechSmith®'s SnagIt®, which allows you not only to take snapshots of screens, but also to capture images of just parts of them, and then edit the images or add callouts and arrows.

Integrating Tables and Figures with Text

When providing captions for both tables and figures, you number each one in consecutive order and reference them within the document: for example, *see Figure 1: the OP System* or *see Table 2: Printer Settings*. When placing captions for figures, you can opt to put them either at the top, bottom, or side, as long as you do so consistently. As stated earlier, you label tables with their titles at the top. Tables aren't figures, so number them separately: for example, *Table 6-1, Table 6-2*, and so on.

While this approach is traditional for integrating figures and tables within your document, you also have the option of not identifying them by using numbers or captions, but rather placing them in close proximity to the text where you mention them.

Color

Like graphic elements, color works to highlight or emphasize parts of text and to orient the reader to different parts of a document. While color provides emphasis and orientation for both print and Internet documents, different types of colors are used for each.

Using Color Effectively

Use color for a definite purpose—not just for the sake of having color. Some purposes for adding color include orienting the reader to where he or she is in a document, separating information for the reader, and reproducing images as they actually appear.

Orientating

If sections of your document are color-coded, readers can find what they are looking for with less difficulty. You can apply color to footers or headers to distinguish chapters or different level headings. In each case, the reader becomes oriented, whether consciously or not, to the system that you want the colors to represent and thus can find sections more easily.

Separating Information

Color can also serve to separate information within your text. For example, you can put colored annotations in a margin, so that the reader can distinguish them from the main text. You can also create side bars with added information and put them in boxes with a tinted background. In both instances, the color separates essential from ancillary information.

Reproducing Images as They Appear

Colored pie charts, photos, and drawings are just some of the many graphics that add a sense of verisimilitude to what they represent. If you reproduce your graphics in color rather than grayscale, they will appear more vivid and realistic.

Types

Print

Colors for print media are called "subtractive," because if you subtract red, green, or blue from white (the combination of all colors), you get cyan, magenta, and yellow, which when combined with black (the absence of light) make up what are called "process" colors. Using these four colors for printing is more expensive than using just one or two colors. Adding a fifth "spot" color, of course, further increases the cost.

CMYK: As stated earlier, the four colors that make up processing for print documents are **cyan (C)**, **magenta (M)**, **yellow (Y)**, and **black (K)**. These colors when mixed in varying proportions become the basis for all colored inks used in printing. In color charts, they are listed in different combinations or percentages and are identified by numbers listed in several systems, the main one being the PANTONE MATCHING SYSTEM®.

Most page layout programs such as FrameMaker® and Adobe® InDesign® allow you to set colors for your documents according to this system or another one such as TRUMATCH®. The benefit of using one of these standard systems is that by designating a numbered color for an image, you get that exact color from whoever prints your document anywhere throughout the world.

Two-Color Versus Four-Color: If you want to use color but have a small budget, you might consider using just two colors—black and another one, usually cyan or magenta. Adding a color other than black to headers or rules enhances a document in most cases and can serve, just as the addition of four colors does, to emphasize text and orientate the reader. Traditionally, using two colors has been less expensive than four, but now with digital printing, the difference is not always obvious and can depend on the amount of pages and work involved in preparing a document for printing.

Spot Color and Tints: However, if cost is not a major concern, in addition to using four colors you might consider adding a fifth or "spot" color to create a special effect. Often this spot color appears as metallic (silver or gold), and will cost extra. A less expensive option is to create a tint of a process color, which is less than 100% of the original color. You can select both options with InDesign® and other page layout programs.

Online

For documents that will appear on screens, you want to use additive colors—colors that, when 100% of each is added, create white. These colors are **RGB: red (R), green (G),** and **blue (B),** and you can select them as you select CMYK in most page layout programs.

Checklist

	Do You Know	Yes	No
1.	when to use tables?		
2.	the different types and parts of tables?		
3.	what graphics are considered figures?		
4.	when to use line charts?		
5.	what types of bar charts exist?		
6.	when to use a pie chart?		
7.	what to keep in mind when creating a pie chart?		
8.	when to use drawings and diagrams?		
9.	when to use photos?		
10.	what to keep in mind when integrating tables, figures, and photos within text?		
11	three uses of color?		
12.	what CMYK and RGB are and their uses?		
13.	what the Pantone Matching System® is?		
14.	what a spot color is?		
15.	why you might use two colors, and what colors they usually are?		

Exercises

1. Match the parts of a table (title, row, column, stub-head, or note) to each of the letters in the table below:

a.				
	b.			
d.	**d.**	**d.**	**d.**	**d.**
c.				
c.				
c.				

e. amounts are in US dollars

2. Evaluate the following charts (Figures 6-10 and 6-11). Is the information clear? Might the charts be improved or another type of chart be used to convey the same information more effectively?

a.

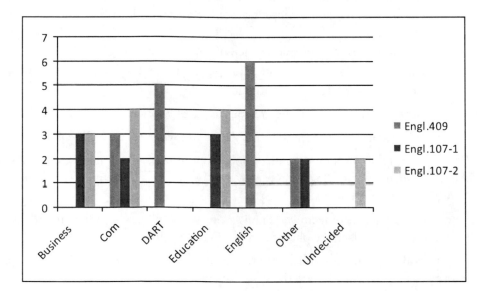

Figure 6-10: Students in an instructor's three classes classified by majors.

b.

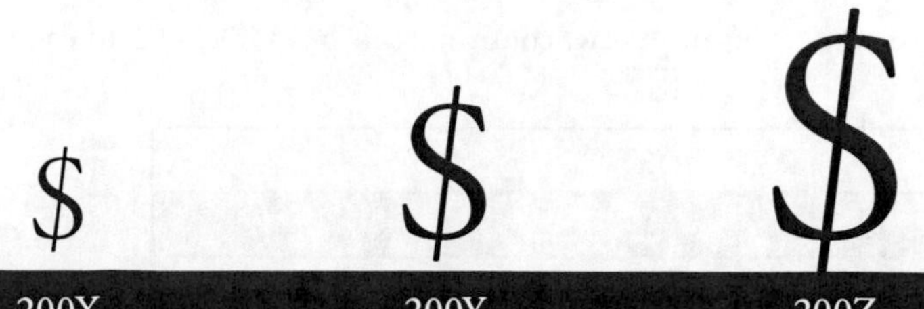

200X	200Y	200Z
$40K	$50K	$60K

Figure 6-11: Revenue generated through sales over three years.

3. Tell what graphic you would use for each of the following:
   a. Monthly sales figures for an automobile dealership for each of its three sales divisions for one year.
   b. Comparison of Republicans and Democrats who voted in the last election based on ages: 18-25, 26-35, 36-49, 50-62, 63-75, 75+.
   c. Percentage of Democrats who voted in the last election who belonged to one of the following religious groups: Catholic, Protestant, Jewish, Moslem, Hindu, or others.
   d. Insurance rates for men and women of various ages.
   e. Image of a newly released printer to appear on the front panel of a brochure.
   f. Instructions for setting up a printer.

4. Create charts for a, b, and c in the previous exercise.

5. Give specific examples of when to use CMYK, and when to use RGB colors.

6. Examine several logos for corporate or non-profit organizations that use color. Are the colors the same whenever these logos appear? Why? Can you research and identify the Pantone® color(s) for one of them?

7. Find several examples of technical publications where photos appear, and evaluate their use. Could drawings replace the photos and be effective? Discuss why or why not.

8. Find several examples of publications where color serves to orient a reader and separate information. How effective are these uses of color?

9. Revise an earlier assignment submitted for this course or another course by adding at least one figure and using color.

10. Devise three figures or tables that you can incorporate into assignments for this course.

CHAPTER SEVEN

Designing Pages or Screens and Using Graphic Elements

This chapter looks at five principles of design—unity, repetition, contrast, alignment, and proximity—especially as they apply to page layout. Discussion of layout includes dividing areas into columns and frames as well as the use of spacing—between characters and words, between lines in a paragraph, and between text and graphics. This chapter also examines the different types of graphic elements—logos, boxes, borders, rules, headers, and bullets—that a writer can use to produce attractive and readable documents. It focuses not only on how to create them electronically, but also how to use them effectively.

Objectives

- ✓ Understand the basic principles of design as they apply to the layout of pages and screens.
- ✓ Learn how to create pages with multiple columns.
- ✓ Understand the use of spacing between characters, words, and lines.
- ✓ Learn about the different types of graphic elements.
- ✓ Understand the integration of graphic elements within text.

Principles of Design

While there are many principles of design, the following five are basic ones that can help you design attractive pages and screens.

Unity

With unity, all the elements on the page—text and graphics or graphic elements—are integrated and brought together. This principle, while readily understood, is sometimes difficult to explain. You know when it exists in page design because all elements fit together.

Repetition

Repeating the same element or pattern on a page helps to unify a design and makes it easier for a reader to grasp what is being communicated. These repetitive elements can be fonts, lines, colors, or objects. Also, their sizes and their placement on a page can be repeated. A successfully designed page or screen will always have some repetition within it. Figure 7-1, for instance, explains how to

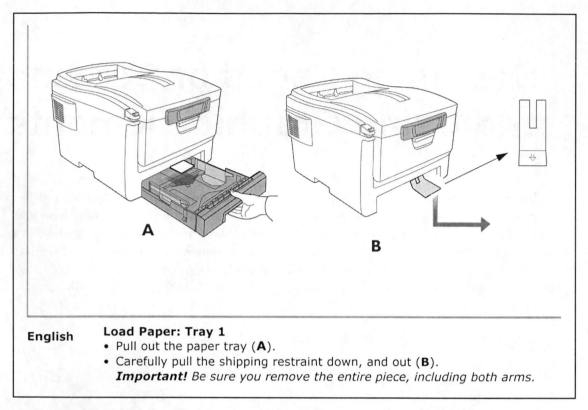

English | **Load Paper: Tray 1**
• Pull out the paper tray (**A**).
• Carefully pull the shipping restraint down, and out (**B**).
***Important!** Be sure you remove the entire piece, including both arms.*

Figure 7-1: Instructions for loading paper into a printer's tray. *Reprinted courtesy of Oki Data Americas, Inc.*

load paper into a printer's tray. Here the two drawings are repeated in varying forms, and the steps themselves each begin with the present part of a verb.

Contrast

Contrasting sizes of type and objects works to create an attractive layout. Other elements that contrast effectively are colors, column widths, and rules. Effective design usually involves some form of contrast, for example, light and dark images or tight clusters of black text within loose areas of white space. Figure 7-1 shows contrast between type weights: bolding is used for the callouts on the drawings, and the normal weight is used for the type in the instructions. Also, having two different drawings provides contrast.

Alignment

Aligning text and parts of illustrations works to create an effective page design. In word processing as well as page layout programs, you can align text either right, left, center, or justified easily by highlighting the text and clicking on an icon that shows the alignment that you want. By aligning your information, you allow your reader to follow more easily what you're saying or showing. Figure 7-2 provides an illustration where all the pieces of packaging foam are aligned vertically above and below the printer so that the reader can quickly visualize how to remove them.

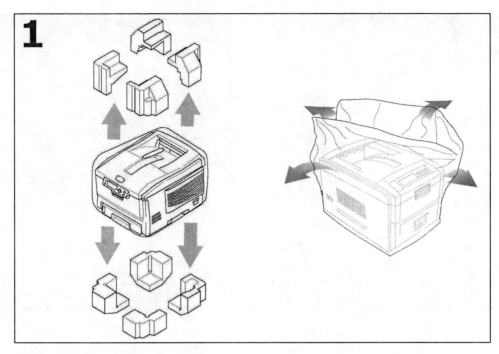

Figure 7-2: Illustration from instructions on setting up a printer showing alignment and proximity. *Reprinted courtesy of Oki Data Americas, Inc.*

Proximity

Putting similar types of information and graphics together works to organize information and to enhance its readability within both print documents and Internet screens. With proximity, you show the reader what you want him or her to see as a unit—what types of information he or she should digest as a whole in comprehending the meaning of your text or illustration. For example, Figure 7-2 shows all the packaging pieces for a printer placed together on the left side of the illustration with the covering on the right side, so the reader looks at these pieces as a unit.

Aspects of Design

Columns

As part of page design, you want to consider where to place your text so that it's most readable and so that it works with any graphics that you might include on the page. If you run your text as lines across the page and thus have **one column**, your text may be difficult to read if there is insufficient space between these lines. However, **multiple columns** need less leading between lines for the text to be easily read.

Multiple columns can be created in Microsoft® Word by using the "Columns" command in the Page Layout tab (Figure 7-3).

While doing so will allow you to put your text in several multiple columns with designated widths, your document might appear attractive if you break up the **gutters** (the spaces between

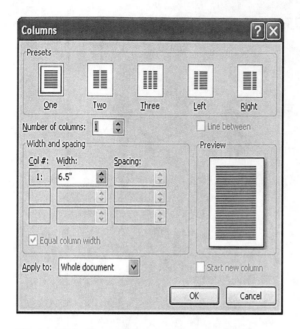

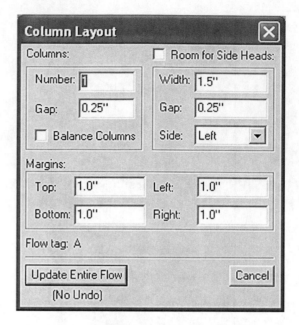

Figure 7-3: Formatting text in columns in Microsoft® Word. *Reprinted with permission from Microsoft Corporation.*

Figure 7-4: Formatting columns in Adobe® FrameMaker®. *Reprinted with permission from Adobe Systems Incorporated.*

columns) also. In newsletters, brochures, and overviews, you can place photos or charts across the gutters so that the gutters don't extend without interruption all the way down the page from top to bottom.

In most page layout programs, not only can you create multiple columns, but you can also adjust gutters. In Adobe® FrameMaker® for example, you can create the number of columns you want and determine the width of the gutter or "gap" between them. Furthermore, you can place your headers either on the side of your text or above it. Figure 7-4 shows a pull-down menu with these various options.

Spacing

Spacing is a key to effective layout or design of both pages and screens. Usually one thinks of space around text and graphics when discussing spacing, but the spacing between words and letters as well as that between lines is also an important part of design.

Kerning and Tracking

Type can be extended or condensed manually or automatically by adjusting the kerning or tracking functions of page layout programs. While "kerning" usually refers to the spacing between a pair of letters, and "tracking" refers to the spacing between all the letters, often the two are used interchangeably or referred to just as *tracking*. Figure 7-5 provides an example of normally spaced type and type that can be extended through tracking.

Technical Communication
(normal spacing)

Technical Communication
(extended spacing)

Figure 7-5: Type extended by tracking.

Leading

Leading refers to the space between lines. The word "leading" comes from the blocks of lead that were originally placed between lines of type in the early days of printing. For type formatted in one column, the leading needs to be wide enough so that the reader's eye can travel across the entire width of the page to read a line and still comprehend its meaning without difficulty. With type formatted in two and three columns, the reader's eye travels a shorter distance across the page to read a line, so the leading can be less. Thus, the smaller the column width, the smaller the amount of leading you'll need. Figure 7-6 provides examples of text in a one-column format and text in a three-column format with different leading.

White space

Pages or screens with sufficient white space surrounding the text and graphics are typical of modern layouts. Compare the crowded page of text shown earlier in Figure 1-1 (Chapter One) from a technical document created in the 1980s with a page created more recently (Figure 7-7) to see how the additional space around the text increases readability.

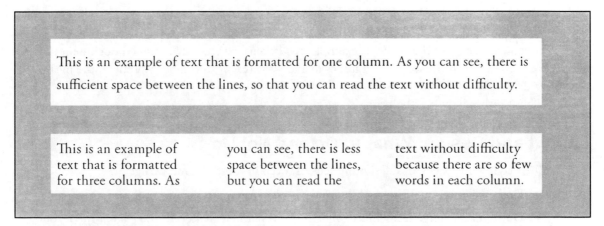

This is an example of text that is formatted for one column. As you can see, there is sufficient space between the lines, so that you can read the text without difficulty.

This is an example of text that is formatted for three columns. As

you can see, there is less space between the lines, but you can read the

text without difficulty because there are so few words in each column.

Figure 7-6: Leading between texts formatted for one column and three columns.

Acies/One Notices Configuration

In the Acies/One property and casualty insurance processing system, the Notices function works with both the Policy and Billing modules. In the Policy module, Notices is used to generate:

- Notices of Conditional Renewal
- Notices of Intent to Reinstate (NOIC Rescission)
- Notices of Intent to Cancel for Underwriting Reasons or Agency Bill Nonpayment
- Notices of Non-Renewal

In the Billing module, Notices is used to generate:

- Notices of Intent to Cancel for Direct Bill Nonpayment
- Cancellations for Direct Bill Nonpayment

Similar to other functions in Acies/One, the setup and configuration process for Notices spans several System Tables and other system administration modules.
System Tables:

- Policy Action Type
- Underwriting Action
- Transaction Code
- Reason Code
- Reason Code Link

System Settings:

- Notices Configuration

Figure 7-7: Introductory page from Acies/One Notices Implementation Guide. *Reprinted courtesy of IDP, Inc.*

As you can see from the previous example, the layout in a recently designed document uses white space around both the text and graphics to allow the reader to fully comprehend the meaning of the words. Notice also that the white space in this example is not locked within one area or "trapped," but rather flows off the page.

Layout and Design: What to Do and Not Do

When laying out pages and designing screens, keep in mind the following:

- **Do** keep all elements in modules so that you can move them around.
- **Do** select one large element as *dominant*, and put it at the top of the page. (The reader will focus first on what the writer considers the most important part of the page.)

- **Do** use sufficient spacing between letters and words, and use tracking, if necessary, to keep them together.

- **Do** have sufficient leading between lines of text formatted in one column.

- **Don't** use excessive leading for lines of text formatted in two or three columns.

- **Don't** trap white space within text: have it flow off the edges of the page.

- **Don't** neglect the smaller points on the page—and especially those elements at the bottom.

Using Software to Lay Out Pages and Screens

There are several programs that can make page layout easier. In professional and technical writing, FrameMaker®, Adobe® InDesign®, and QuarkXPress® are some choices. While all these programs can help you lay out attractive pages for print, many of them also allow you to export these pages as screens. However, if you're laying out screens, you would probably want to use Adobe® Dreamweaver® or Adobe® GoLive®.

Graphic Elements

Uses

Graphic elements are design features that make a page attractive and aid in a reader's comprehension. When designing your pages, you want to use them to enhance the overall readability because they provide needed breaks for readers from viewing large passages of text, they help orient readers to relationships among ideas, and they hold together information in key places.

Types of Elements

Graphic elements include logos, boxes, borders, rules, and typefonts, especially those used for headings.

Logos

Logos consist of either type alone or type with images or shapes that symbolize an organization or product. In professional and technical writing, they can effectively substitute for text to both communicate a message and provide an identity for a particular concept or brand.

Boxes and Borders

Boxes consist of frames around objects and text. They can be created with shadows to appear three-dimensional. Borders with different line widths and a variety of designs can surround boxes and pages. Both boxes and borders help break up pages of text and organize text in such a way as to increase its readability. The "Borders and Shading" menu in Microsoft® Word (Figure 7-8) allows you to create a variety of borders using lines of different widths and provide shadowing or a three-dimensional effect.

Similarly, in FrameMaker® the tool bar allows you to select tools to create shapes, lines, boxes, or shading.

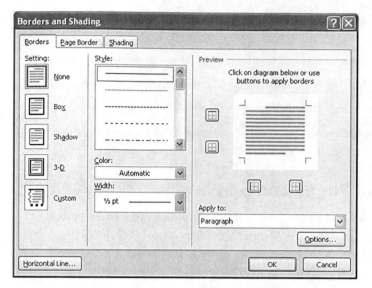

Figure 7-8: Pull-down menu in Microsoft® Word for creating boxes, borders, and shading. *Reprinted with permission from Microsoft Corporation.*

Figure 7-9: FrameMaker® tools palette for creating and modifying rules. *Reprinted with permission from Adobe Systems Incorporated.*

Rules

Rules also help to break up blocks of text and serve to separate different topics. In Microsoft® Word as shown earlier, selecting "Borders" allows you to select line widths of rules. You can also create rules and modify them in Microsoft® Word with drawing tools: the *line* tool itself as well as the *line style* and *line color* tools. In a layout program like FrameMaker®, you can create rules and vary their color and width by using tools on the tool palette (Figure 7-9).

Typefonts

Typefonts serve as graphic elements because they communicate through their size and shapes. For example, Times Roman or New Times Roman, the default fonts in earlier versions of Microsoft® Word, convey a rather conservative image because they are so well known as body fonts and because they are *serif* fonts—those with small feet and lines hanging off their letters —and thus somewhat delicate and traditional. On the other hand, *sans serif* fonts—those without the small feet and lines—convey a more modern, sleek image. If you choose a serif font for your body text, consider using a sans serif font for your headers to create contrast. While Arial or Calibri is a default sans serif font for many programs, there are many other attractive sans serif fonts you can use: for example, Franklin Gothic, Gil Sans, Verdana, or **Impact**. Microsoft® Word and most versions of page layout programs have a wide selection of fonts. However, if you want to use additional fonts, consider searching for them on the Internet where you can purchase them or—in some instances—download them without cost.

Franklin Gothic Demi
Franklin Gothic Medium
Franklin Gothic Heavy

COPPERPLATE GOTHIC LIGHT
COPPERPLATE GOTHIC BOLD

Gils Sans MT
Gills Sans Ultra Bold

Figure 7-10: Differing weights of three type families.

Cambria (Roman or straight)
Cambria (italics or slanted)

Calibri (Roman or straight)
Calibri (italics or slanted)

Figure 7-11: Postures of default fonts in Roman and italics.

Size: While you'll probably use a 9- to 12-point typefont size for your body text, you want to choose a larger one for your first-level headers. Depending on how many different levels of headers you anticipate, you might choose a large size (14- to 16- point) for your title and first-level headers, and use a smaller size (12- to 14- point) for the next level headers. If your headers are the same size as your body text, you will need to change their weight or posture to differentiate them from the text.

Weight: The "weight" of a typefont is categorized by its thickness. Common weights include light, demi, medium, bold, and ultra bold. Figure 7-10 shows several popular weights of different type families.

Posture: You can also vary your typefonts by varying their "postures." Most often you'll use Roman fonts, which are straight fonts, but for different level headers or for emphasis you might use italics or slanted postures. See Figure 7-11 for these different postures.

Other Effects: Other ways you can style typefonts include underlining, shadowing, and CAPITALIZATION. These effects, like the others, emphasize sections of text or serve to provide additional contrast for sub-headers.

Lists and Bullets

Other graphic elements include lists and bullets. Putting a series of items in a list rather than leaving them in a sentence helps the reader to grasp their meaning. Adding bullets can make the items in the lists stand out and appear easier to read.

To distinguish more clearly between different levels for your lists, you can use different types of bullets for each. In Microsoft® Word, as shown in the drop-down menu in Figure 7-12, you can create symbols as bullets, in addition to clear and solid bullets of different sizes and shapes.

Keep in mind the importance of *parallel structure* for all lists. If the first bulleted item is a noun, verb phrase, or a sentence, then the items that follow should be the same grammatical part or structure. Also, with lists, keep in mind your *indentation* of items. For different levels of lists, you can increase your indentations as in the following example, where the second-level items are indented farther to the right than the first-level items:

- Tools
  - o Hammer
  - o Screwdriver
  - o Wrench
- Equipment
  - o Floor polisher
  - o Waxer
  - o Dry vacuum cleaner
- Services
  - o Cleaning
  - o Refinishing
  - o Polishing

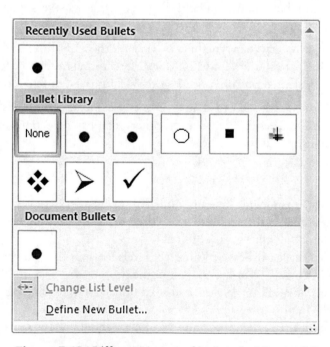

Figure 7-12: Different types of bullets in Microsoft® Word. *Reprinted with permission from Microsoft Corporation.*

Checklist

	Do You Know	Yes	No
1.	five principles of design and how they work to communicate information?		
2.	how to create multiple columns?		
3.	what gutters are in the design of a page?		
4.	the difference between kerning and tracking?		
5.	what leading is and how it affects readability?		
6.	why having sufficient white space is important?		
7.	why you want to place a "dominant" element at the top of a page?		
8.	programs for laying out print pages and for laying out screens?		
9.	names of five graphic elements?		
10.	what boxes, borders, and rules do?		
11	how typefonts work as graphic elements?		
12.	some of the changes that can be made to typefonts to differentiate body text from headers and different levels of headers?		
13.	what other stylistic effects you might use?		
14.	what to keep in mind when creating lists?		
15.	how to create different types of bullets for a list?		

Exercises

1. Which of the five principles of design do you think is most important?

2. Can text serve as a dominant element on a page or screen, or should only a graphic dominate?

3. What do logos consist of, and why are they able to symbolize so effectively an organization or group?

4. How can rules work to enhance readability?

5. What should you keep in mind when using a single or multiple column format?

6. Explain the difference between *white space* and *leading* and between *kerning* and *tracking*.

7. Explain how you can use different typefonts and their family members as graphic elements.

8. What is the advantage of putting items in a list rather than in sentences?

9. Explain the benefit of using bullets and using different types of bullets.

10. Locate two web sites that contain technical information. You can go to those from a previous exercise in an earlier chapter (for example, the Heart Information Network, NASA, or the Government Security Act Regulations in Chapter One) or any others. Write an evaluation of each site by describing how it embodies principles of design (unity, repetition, contrast, alignment, and proximity) and use the following elements:

    - White Space
    - Type (size, font, capitals or lower case, highlighting)
    - Lines/Margins
    - Line lengths
    - Columns
    - Alignment
    - Headings

PART III:
Integrating Text and Graphics

This section focuses on the integration of text and graphics in a variety of technical, business and promotional documents used in written and oral communications.

Chapter Eight
Definitions

Chapter Nine
Descriptions

Chapter Ten
Instructions

Chapter Eleven
Proposals

Chapter Twelve
Reports

Chapter Thirteen
Correspondence

Chapter Fourteen
Guides and Promotional Materials

Chapter Fifteen
Oral Presentations

PART III:
Integrating Text and Graphics

This section focuses on the integration of text and graphics in a variety of technical, business and promotional documents used in written and oral communications.

Chapter Eight

Definitions

Chapter Twelve

Reports

Chapter Nine

Descriptions

Chapter Thirteen

Correspondence

Chapter Ten

Instructions

Chapter Fourteen

Guides and
Promotional Materials

Chapter Eleven

Proposals

Chapter Fifteen

Oral Presentations

CHAPTER EIGHT

Definitions

Definitions appear throughout technical documents. This chapter focuses on the use and types of simple definitions and explains how to expand them through various rhetorical modes. It also looks at integrating graphics to expand definitions visually.

Objectives

- ✓ Understand the use and types of simple definitions in technical communication.
- ✓ Understand the use of expanded definitions as they relate to simple definitions.
- ✓ Learn how to create expanded definitions using several rhetorical methods.
- ✓ Learn to integrate graphics with text in creating definitions.

Definitions

Definitions appear throughout overviews, instructions, and manuals. They can range from a simple parenthetical explanation in a sentence to a paragraph entry in a glossary or to an in-depth discussion of several pages. Whatever the difference in complexity, they all should begin with a simple definition.

Simple Definition

A simple definition explains the term by giving the larger class to which a term belongs and then the differentia that distinguishes it from other members in that class. As shown in Figure 8-1, the *class* is the larger generic group (computers), and the *differentia* is one of the sub-groups that belongs to it.

Uses

Simple definitions are most commonly used in the body of text where the term first appears. If the definition is short, it often follows the term in parentheses; if it's longer, it usually appears in a complete sentence. If the definition is somewhat complex, it can require two or three sentences and appear in a note or glossary of terms. A glossary contains simple definitions of terms that appear frequently in long documents.

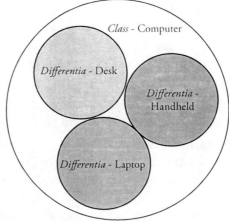

Figure 8-1: Relationships among parts as in a simple definition.

Types

Examples of both types of simple definition follow:

Parenthetical definition: adware (unwanted online advertisements that are downloaded unknowingly onto one's computer)

Sentence definition: Adware is an online advertisement that pops up on a screen and has been downloaded onto a computer without a person's knowledge or consent.

In both examples, the term *adware* is defined first by placing it in a class (for example, *online advertisement*) and then adding a word or phrase that differentiates it from other terms in that class. For the parenthetical definition, the differentia consists of the word *unwanted and* the phrase *downloaded unknowingly onto one's computer.* For the sentence definition, the differentia is the clause *that pops up on a screen and has been downloaded onto a computer without a person's knowledge or consent.*

Expanded Definition

Sometimes you need to expand a definition so that its meaning will be clearer. The uses of expanded definitions and the methods to expand them follow.

Uses

While expanded definitions often appear within manuals and other longer documents, they can appear in short documents as well, especially if the term being defined is complex, and clarity is needed. Sometimes, in fact, expanded definitions occupy a paragraph or several paragraphs, such as the definition of chemotherapy in Figure 8-2. To expand a definition of a term or concept, you need to begin with its sentence definition and then explain it further using several rhetorical methods.

Methods of Expansion

Origin or history: One popular way of clarifying what a term means is to tell its word origin or relate how its present meaning came into being. For example, in Figure 8-2, the writer defines chemotherapy by telling us who first coined the term, who is credited with first using it, and how chemotherapy became an effective treatment during World War II.

Synonyms: You can also explain a term's meaning more fully by providing synonyms for it—especially if you choose those that have meanings similar to but slightly different from the term itself. For example, for *adware*, you might supply words such as *spam* or *junk mail*. Though neither is the same, the connotation conveyed by both of something unwanted and something that was meant to sell a product gives the reader a better understanding of what adware is.

Negation: You can also define a term by explaining what it is not. For example, you might write that *adware is not necessarily a cookie.* Although a "cookie" might be an unwanted program that produces adware on your computer, these terms are not the same.

Comparison-Contrast: Another method is defining a term by comparing and contrasting it with what is similar or dissimilar to it. For example, you might explain the cyclical process of evaporation and condensation by comparing the process to water being sucked up into a dry vacuum machine and then being emptied from it.

Using contrasts, you can also explain a term's meaning by describing how it differs from something seemingly similar. For example, you might explain *reflection*, the means by which light is thrown or bent back, by explaining that it's unlike *refraction*, the means by which light is deflected or turned aside. Actually, the sharper the contrast or more delineated the difference, the clearer the meaning will be of the term you are defining.

Classification: You can expand the definition of a term by classifying its various parts and describing each of them in detail. While not every term lends itself to this approach, if the term represents a whole that can readily be divided into parts, you can clarify this term's meaning for the reader by such classification. For example, you might explain *Judaism* by classifying different types: *orthodox, conservative,* and *reformed.*

Examples: Another way to expand your definition is to use examples—either brief ones that need little elaboration because they are readily recognizable to your reader, or more fully developed case histories that have extensive details. For instance, you might define an *analgesic drug* by using *aspirin* or *ibuprofen* as examples, or you might explain a concept such as *hydroponics* with a detailed example of its use.

Cause-Effect: You can also explain a term's meaning by describing its causes or its effects. You can clarify the meaning of a term such as *dehydration* for someone if you tell what circumstances or what situation brings about this condition. You can also make it clearer if you explain what the effects are. In defining by cause-effect, keep in mind that you are likely to find **immediate** causes and effects as well as more **remote** ones. In the following segment (Figure 8-2), the writer gives an immediate effect of chemotherapy—namely, that it attacks cells—and also gives a remote effect—that it stops cancer from spreading.

Extended Definition of Chemotherapy

[SENTENCE DEFINITION]
Chemotherapy is a form of medical treatment for infection or malignant disease that uses chemical agents that are directed against a specific organism or abnormal cells.

[WORD ORIGIN AND HISTORY]
The term chemotherapy was coined by the Jewish bacteriologist and chemist Paul Ehrlich (1854-1915). The first people to use chemotherapy were the ancient Egyptians, who used drugs to treat cancer. However, effective use of chemotherapy started during World War II, when scientists discovered that mustard gas destroys the cells of the lymphatic system. Once researchers learned the nature of these chemicals, they began to wonder if chemicals could also be used to selectively destroy unwanted cells. This research led to the development of drugs that could combat cancer of the lymphatic system. Soon after, drugs were developed to treat many different types of cancer. Currently, there are approximately 50 different drugs that treat cancer.

[CAUSE-EFFECT]
While chemotherapy is designed to attack foreign organisms or abnormal cells, it also attacks normal cells. Chemotherapy is most often used in the treatment of cancer. Some drugs are able either to stop cancer cells from spreading or even to prevent the start of cancer. The drugs that are used in chemotherapy, called chemotherapeutic agents, destroy cells by interfering with the normal, life-sustaining functions of individual cells. Chemotherapeutic drugs work in one of the following ways:

Figure 8-2: Segment of a definition expanded by several methods.

Drafting a Definition: What to Do and Not Do

Keep in mind the following guidelines in creating your definition—whether a simple or expanded one:

- **Avoid defining a term circularly.** In other words, don't define a term by using a variation of it: for example, a noun or adjective form of the same term: for example, *Combustion is the state that occurs when something is combustible*.

- **If you use a linking verb between the term you are defining and its meaning, use the same part of speech for the term and its meaning.** Linking verbs like *is, are, was,* or *were* act like equal signs in mathematical equations. Therefore, the parts of speech should be the same on each side of them (for example, noun or pronoun= noun or pronoun) and not different (noun or pronoun ≠adverb or conjunction). For example, write *Chemotherapy is a process* rather than *Chemotherapy is when*. If you can't use the same parts of speech, then change the linking verb to another type of verb: for example, *Chemotherapy occurs when…*

- **Don't use words in the definition that your reader is unlikely to know.** Use only words in your explanation that your reader would understand; otherwise you'll need to define additional words.

Designing Pages and Using Graphics

When you publish your definition of a concept or term, consider using a graphic that illustrates it. While some concepts are too abstract to illustrate, some can be explained with symbols or icons often associated with them. Obvious examples would be a dove for peace or a cross for Christianity, since these symbols connote meanings that most readers readily understand.

Keep in mind that if you expand your definition, you might supply headers to distinguish its parts and make its meaning clearer. Remember also that you always need to consider your audience and the required level of formality when selecting graphics and designing pages or screens.

Checklist

	Do You Know	Yes	No
1.	what the class and differentia are in a simple definition?		
2.	where you might use or see a simple definition?		
3.	what format a simple definition can take?		
4.	why you might put a definition in parentheses?		
5.	where you might use an expanded definition?		
6.	several ways to expand a simple definition?		

	Do You Know	Yes	No
7.	what is meant by having a circular definition?		
8.	why you don't want to use words in a definition that would be unfamiliar to your reader?		
9.	why a familiar graphic might help define an abstract term?		
10.	how headers might be used in a complex expanded definition?		

Exercises

1. Explain why definitions are important, and where you might find them in professional and technical communication.

2. Explain the difference between a <u>simple</u> definition and a <u>sentence</u> definition.

3. Name three ways you can expand a sentence definition.

4. What is a circular definition?

5. What is wrong with the following definition? *Telecommuting is when an employee works from his or her home using the Internet or other telecommunication devices.*

6. Why should you carefully consider your audience when selecting graphics for a definition?

7. Write a sentence definition of one of the following terms, naming the class to which it belongs and the features or characteristics that differentiate it:

humidity	computer platform
socialism	defragmentation
bulimia	firewall
hydroponics	cholesterol
computer interface	spyware

8. Expand a sentence definition created in the previous exercise to approximately 400 words, using at least three different methods discussed in this chapter (origin and history, negation, classification, comparison and contrast, and examples).

CHAPTER NINE

Descriptions

Like definitions, descriptions appear throughout technical documents. This chapter focuses on different types of descriptions and how to structure descriptions of objects or mechanisms and processes. It also looks at integrating them with graphics, callouts, and headers.

Objectives

- ✓ Understand the different types of descriptions.

- ✓ Learn how to structure a description of an object or mechanism.

- ✓ Learn how to structure a description of a process.

- ✓ Understand the different ways to lay out text and graphics for pages or screens.

- ✓ Learn about using callouts and headers with technical descriptions.

Types and Components

Descriptions appear in many types of documents—often in overviews of products that appear in brochures and manuals. These can be **general** descriptions of generic items or **particular** ones of specific brands or products. An example of a general description would be one that describes an MP3 player, and an example of a specific description would be one about an Apple iPod.

Two basic types of descriptions—of **objects** or **mechanisms** and of **processes**—have similar parts: an **introduction**, a list of **parts or stages** with explanations of each, and a **conclusion** that usually describes the integration of the parts or stages.

Description of an Object or a Mechanism

Introduction

You might start an introduction with a sentence that explains what the object or mechanism is, and then proceed with a more detailed explanation of what is to follow—your listing of parts, for example, and a statement about how they are integrated. Figures 9-1 provides an example of an introduction from a student's description of an MP3 player.

Parts

After providing an overview of the object or mechanism, you can describe each part in a separate paragraph with a header to identify it. You can begin with details about each part's size, shape, or other characteristics; or describe its use in relation to the other parts. Whatever you do, you want to be consistent in what you say about each one. For example, if you describe one knob

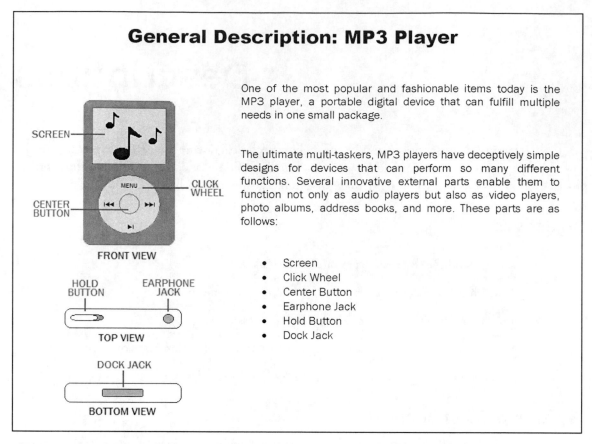

Figure 9-1: Introduction to a general description of an MP3 player.

of a device as 2 inches in diameter and red in color, you also want to give the measurements for the other knob and identify its color. Use callouts to label each part with a name that matches the name in your description, as shown in Figure 9-2. (For a discussion of "Callouts," see Chapter Five and the next to the last paragraph in this chapter.)

Conclusion

After describing the parts of an object or mechanism, you need to conclude with a brief summary of what you've written and perhaps a statement about the function of the object itself (Figures 9-3).

SCREEN
    Because most MP3 players can function as mini video players as well as music players, their crystal-clear screens are a crucial feature – and a main selling point – of the devices.
    While older MP3 players were designed with a two-tone black and white screen, most new models have slightly larger, full color screens. This small improvement not only makes the device's menus more attractive, it allows users to watch TV shows and movies on these portable devices.

SCREEN

CLICK WHEEL
    Perhaps the most innovative feature of the MP3 player is the Click Wheel. The Click Wheel allows users to scroll through menus, fast-forward through songs, and adjust the volume. The Click Wheel is touch-sensitive, so a brush of the fingertip is all users need to find their favorite song, play a video, or adjust volume.

CENTER BUTTON
    Pressing the Center Button executes a command. It has the same effect as clicking on something with a mouse or pressing the Enter key on a keyboard. When navigating through menus with the Click Wheel, pressing the Center Button will allow the user to make a selection.

CLICK WHEEL

EARPHONE JACK
    On the top of an MP3 player, there is a small hole to insert earphones. Most MP3 players are packaged with a set of tiny earphones called "earbuds" that are like miniaturized speakers that users place directly in their ears. However, any standard set of earphones will work with this jack.

HOLD BUTTON
    The Hold Button is a convenient switch on the top of an MP3 player that allows a user to basically disable all other buttons on the device. This is especially useful if someone is carrying an MP3 player in a purse or backpack. When the Hold Button is on, the device will not accidentally turn on if another item in the bag pushes against its buttons. For students, this switch ensures that music will not suddenly come blaring out of their MP3 players in the middle of class if the contents of their bags shift around.

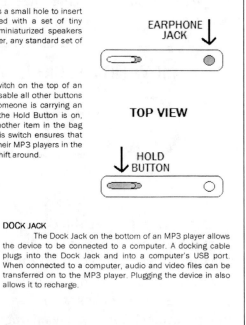

DOCK JACK
    The Dock Jack on the bottom of an MP3 player allows the device to be connected to a computer. A docking cable plugs into the Dock Jack and into a computer's USB port. When connected to a computer, audio and video files can be transferred on to the MP3 player. Plugging the device in also allows it to recharge.

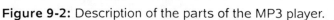

Figure 9-2: Description of the parts of the MP3 player.

MP3 players have not only revolutionized the way people listen to music, but have raised the bar for portable entertainment devices all over the world. These tiny devices have all of the features necessary to keep even the weariest of travelers entertained for hours and the most frazzled businessperson sane and organized. MP3 players are pocket-sized powerhouses that will be a gadgetry staple for years to come.

Figure 9-3: Conclusion in a description of an MP3 player.

Description of a Process

Introduction

Describing a process is similar to describing an object or mechanism, except that instead of describing parts and explaining their interaction, you describe the stages of the process. In such a description you want to begin with an **overview of the process** and proceed to provide a brief **summary of the stages** and any other information that is needed to understand it. Figure 9-4 introduces the reader to a description of the process of putting together a college newspaper.

Stages

Detailed descriptions of stages appear in separate paragraphs and are separated by headers that use nouns or gerunds to identify each of them. For example, gerunds such as *taking, removing, inserting,* or *connecting* indicate the actions that you're describing. Because you're describing rather than instructing, you want to focus on the stage itself rather than on the person involved in that part of the process. Figure 9-5 describes five stages in the process of publishing articles in a weekly college newspaper.

How an Article is Written and Published in the

COLLEGIAN

est. 1931

With a newspaper like the *Collegian*, articles are constantly being planned. Sometimes they can be a month or two in the works; sometimes they are written on the spot Sunday evening. But, usually, there is about a week or two between the articles' conception and their publication. The process of publishing articles in a college newspaper consists of five main stages: pitching an article, meeting with an editor, research/writing, laying out sections, and printing.

Figure 9-4: Introduction for a description of a process.

1. **Pitching an Article:** Articles can be conceived (or "pitched") in one of two ways.

    A writer can develop an idea for a news piece, review, or interview on his or her own and, depending on the type of article, pitch it to an editor. For example, Nick Adams maintains a music blog. Every week, he pitches an album review to the Entertainment section editor.

    On the flipside of that, a section editor, or even the editor-in-chief, can find a story that requires a writer and shop it around to the *Collegian* staff. When an important news story breaks, News section editor Eva Brown might pitch it to her staff to see who is best qualified and interested.

2. **Meeting with the Editor:** Meetings are held every Thursday during the free period. The articles that have been planned so far for the following week's issue are reviewed. Sometimes, articles designated for a particular section, like Commentary, might be better placed in Entertainment, so pieces can be swapped. If an article is strongly opposed by the editorial board, perhaps because of a lack of timeliness or an overabundance of potential controversy, it might also be cut from the paper altogether.

3. **Researching/Writing:** Once an article is approved, the appropriate research needs to be done. For an interview/news piece, this stage involves drafting pertinent questions, setting up an interview time, and then conducting that interview. For an art piece (painting, music, film, or literature), this stage means spending time with the subject so as to develop a clear opinion on it.

    But regardless of the article type, some fact finding needs to take place. The names of people must be found: those involved in the basketball team's big win, those in a band being reviewed, or those traveling to Vietnam over spring break, for instance.

Figure 9-5: Stages in the process of publishing articles in a college newspaper.

4. Laying Out Sections: Articles are due to the *Collegian* editors on Sunday. Then on Monday and Tuesday, the sections are designed and laid out by each editor. Articles are adjusted; and the order, column number and length, headlines, and accompanying pictures are selected. After each page's rough draft is finished, it's printed and sent to the copy editors.

Then the necessary changes are made, and a second copy is printed. Editing this time consists of correcting punctuation and mechanical errors, which should be minimal – two or three per page at most. While this is going on, all of the images for the issue are converted to grayscale or, if an image is going on a color page, to cyan/yellow/magenta/black (CYMK).

Once the photos and final edits are finished, they are saved to each editor's section in the "final" folder.

5. Printing: Once all of the pages are finalized, they are converted into a Portable Document Format (PDF) and e-mailed to the publisher for printing. Approximately 1,000 copies per issue in an 11.25 x 17 format are printed with some sections printed in black-and-white and others printed in color based on the editor-in-chief's specifications.

Figure 9-5: Stages in the process of publishing articles in a college newspaper (*continued*).

Conclusion

Like the conclusion of a description of an object or mechanism, the conclusion of a description of a process ties the earlier short descriptions together or tells something about what happens when the process is completed. Figure 9-6 shows the one-sentence conclusion for the process described earlier.

The following morning, the *Collegian* is distributed around campus with the other newspapers, like *The Daily News*, *The Inquirer*, and many more; at which point the articles are available to be read by students and faculty.

Figure 9-6: Conclusion in the process of publishing articles in a college newspaper.

Designing Pages and Using Graphics

Layout

When laying out pages of your description, place your graphic close to the section of text that it illustrates. You can use the method shown in this book of referencing the figure by number within the text and then identifying it with the same number in a caption at the bottom of the figure itself (you can also place figure captions at the top or sides). Or if there is sufficient space to place the figure directly next to the text, you can do so without any accompanying captions. The figures in the descriptions of the MP3 player and the process of publishing articles in a college newspaper are both examples of such a layout.

Graphics

A variety of graphics exists to illustrate your technical descriptions. While professional photos give an overview of an object or process, if you're describing parts of an object or mechanism, you'll probably use a drawing so that the parts are readily seen without the distraction of any additional items that would appear if you used a photo. For basic illustrations, consider using some simple pieces from clip art, or consider creating some sketches using Adobe® Illustrator® or another drawing tool.

Callouts

For drawings, you might need callouts—identification markers that "call out" the name of the part you're showing. In creating "callouts," always place them on a horizontal plane (not vertical or diagonal), and keep the text for them short. You can put them next to the part, above or below—as long as there is sufficient space (see Figures 9-1 and 9-2 for examples of these placements). If there is insufficient space for the complete names of the parts being identified, identify them with letters or numbers, and place a listing nearby where the reader can readily see the name of the part that each letter or number represents.

Headers

In addition to supplying callouts for illustrations, you might want to provide headers that clearly identify each of your sections or stages. If any sub-divisions exist, you might want to provide additional headers (second and third-level ones) with typefonts that differ in size, posture, or weight from the main or first-level ones so that the reader can easily recognize each level by the font size and its appearance—for example whether it's italics or bold. Always make sure that all headers for each level are the same grammatical parts or structures: for example, all nouns, all noun phrases, or all gerunds.

Checklist

	Do You Know	Yes	No
1.	the difference between a particular and a general description?		
2.	how to structure a description of an object or mechanism?		
3.	what callouts are and where you should place them on an illustration?		
4.	why it's important that the listing of parts in a description of an object matches the order of callouts in its illustration?		
5.	what to include in the introduction of a description of a process?		
6.	why the names of the stages of a process you're describing are often expressed as gerunds?		
7.	what the concluding sentence or paragraph does in a description?		
8.	what options exist for integrating text with illustrations in a description?		
9.	where to place graphics within the description?		
10.	why drawings rather than photographs usually accompany text describing parts of an object or mechanism?		

Exercises

1. Explain the difference between an object and a mechanism.
2. Explain how the structure of a description of an object or mechanism differs from the description of a process.
3. What are your options in using figures to illustrate your descriptions of parts or stages? When do they need to be referenced in your text?
4. What is the difference between captions and callouts?
5. If you have sub-headers, how might you differentiate them from the main headers?

6. Select one of the following (A) objects/mechanisms or (B) processes, and list four or five parts or stages that you think you would need to explain in a description of it:

   A. **Objects or Mechanisms**

       digital camera

       toaster

       air conditioner

       hair dryer

       hedge trimmer

       electric can opener

       microwave oven

       MP3 player

       garage door opener

       CD burner

       DVD player

       scanner

   B. **Processes**

       how paper is made

       how an ink jet or a laser printer works

       how an offset printing press operates

       how a piano is tuned

       how a runner prepares for a marathon

7. Select an object or process from Exercise 6, and sketch three or four illustrations that you think would be needed for a description of it. Explain where you would place each.

8. Prepare a two-page description of one of the objects or processes in Exercise 6, and integrate at least two illustrations within the text.

CHAPTER TEN

Instructions

Instructions are used for assembling, installing, and operating equipment as well as for completing procedures within organizations. In many ways, instructions are what we think of as technical communication. When most effective, they combine the integration of both text and graphics in an almost seamless structure.

Objectives

✓ Analyze processes to determine which type of instructions to devise.

✓ Learn to write instructions that effectively lead a reader through a process to its completion.

✓ Learn to integrate graphics with text so that they complement each other and effectively communicate the instructions.

✓ Know when to use notes, warnings, cautions, and dangers and how to place them appropriately.

Types

Assembly

Some products need to be assembled before you can use them. In assembling a product, sometimes you can put together its parts in any order, while at other times you have to follow a very specific order. Whatever the case, when creating assembly instructions, keep in mind the importance of clearly and legibly labeling each part and of indicating with graphics the action involved. For example, Figure 10-1 shows set-up instructions for a printer where all parts are clearly labeled; and Figure 10-2 shows arrows indicating the actions required for steps in this process.

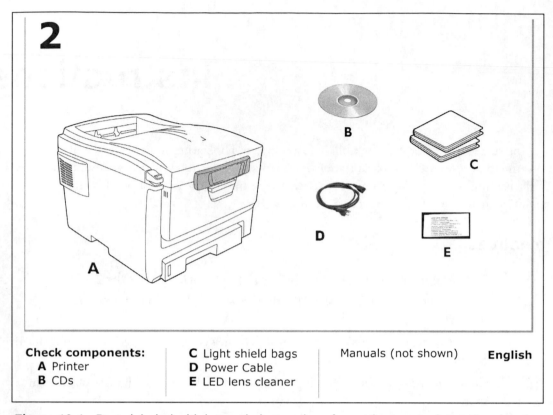

Check components:
A Printer
B CDs

C Light shield bags
D Power Cable
E LED lens cleaner

Manuals (not shown)   **English**

Figure 10-1: Parts labeled with letters in instructions for setting up a printer. *Reprinted courtesy of Oki Data Americas, Inc.*

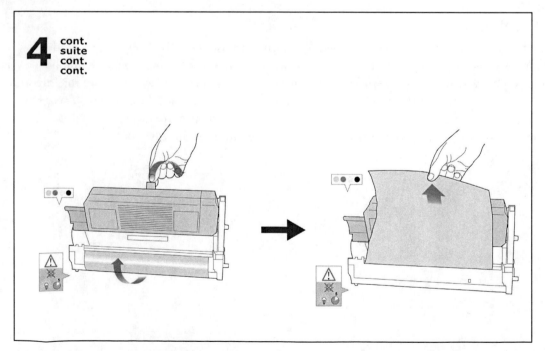

Figure 10-2: Arrows indicating the required action in instruction for setting up a printer. *Reprinted courtesy of Oki Data Americas, Inc.*

Installation

Products often need installation instructions. Usually, they consist of a list of necessary materials or tools, followed by the directions. Graphics especially help to explain an installation process and are useful in showing where to put parts together and what the end result should look like. Figure 10-3 shows instructions for installing a door where drawings with callouts explain each step.

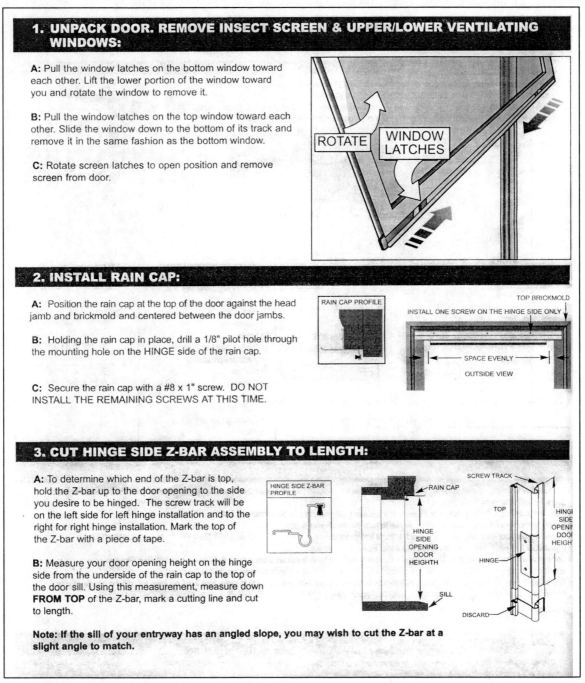

1. UNPACK DOOR. REMOVE INSECT SCREEN & UPPER/LOWER VENTILATING WINDOWS:

A: Pull the window latches on the bottom window toward each other. Lift the lower portion of the window toward you and rotate the window to remove it.

B: Pull the window latches on the top window toward each other. Slide the window down to the bottom of its track and remove it in the same fashion as the bottom window.

C: Rotate screen latches to open position and remove screen from door.

ROTATE

WINDOW LATCHES

2. INSTALL RAIN CAP:

A: Position the rain cap at the top of the door against the head jamb and brickmold and centered between the door jambs.

B: Holding the rain cap in place, drill a 1/8" pilot hole through the mounting hole on the HINGE side of the rain cap.

C: Secure the rain cap with a #8 x 1" screw. DO NOT INSTALL THE REMAINING SCREWS AT THIS TIME.

RAIN CAP PROFILE

TOP BRICKMOLD
INSTALL ONE SCREW ON THE HINGE SIDE ONLY
SPACE EVENLY
OUTSIDE VIEW

3. CUT HINGE SIDE Z-BAR ASSEMBLY TO LENGTH:

A: To determine which end of the Z-bar is top, hold the Z-bar up to the door opening to the side you desire to be hinged. The screw track will be on the left side for left hinge installation and to the right for right hinge installation. Mark the top of the Z-bar with a piece of tape.

B: Measure your door opening height on the hinge side from the underside of the rain cap to the top of the door sill. Using this measurement, measure down **FROM TOP** of the Z-bar, mark a cutting line and cut to length.

Note: If the sill of your entryway has an angled slope, you may wish to cut the Z-bar at a slight angle to match.

HINGE SIDE Z-BAR PROFILE

RAIN CAP

HINGE SIDE OPENING DOOR HEIGTH

SILL

SCREW TRACK

TOP

HINGE

DISCARD

HINGE SIDE OPENING DOOR HEIGTH

Figure 10-3: Installation of a storm door. *Reprinted courtesy of EMCO Enterprises.*

Operation

Operation instructions abound within technical communication. To create them, you yourself must first understand how to use the equipment or mechanism and then determine the best way to explain this process to someone else. As with preparing other types of instructions, you need to analyze your audience fully so that you choose language and graphics appropriate for them. Figure 10-4, for example, shows operating instructions with simple graphics, color, and "Handy Hints" designed for an average consumer who has purchased an inkjet printer.

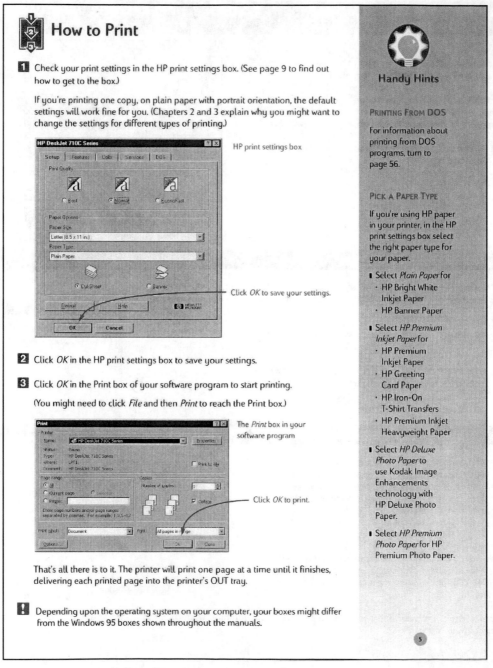

Figure10-4: Instructions for printing. *Reprinted courtesy of Hewlett Packard, Inc.*

Procedure

Unlike other types of instructions, procedures are not always associated with tangible products. They can vary from directives for evacuating people during a natural disaster, to a plan for losing a desired amount of body fat, to steps to follow when hiring new employees or filing a complaint about discrimination. In all cases, the writer of a procedure must take into account the people who will be reading it, their familiarity with the topic, and the amount of time they will spend completing it. Figure 10-5, for example, shows a procedure that an undergraduate wrote for other undergraduates who want to apply to medical school. In it, she alludes to courses and activities she knows other students would be familiar with.

How to Apply to Medical School:
A Step-by-Step Guide for Undergraduate Biology at Larson University

Introduction

Applying to medical school is a complicated process. In order to become a qualified applicant for medical school, there is a lot of work that must be done throughout your four years of undergraduate study. If you follow this guide, you will be well prepared to apply to and attend the medical school of your choice.

Freshman Year

1. Take introductory biology courses (Bio 210 and Bio 220) and general chemistry classes (Chm 111 and Chm 112).

2. Get involved in an on-campus activity. Examples of extracurricular activities include community service, sports, yearbook, newspaper, biology clubs, and fraternities or sororities.

3. Begin volunteering at a local hospital. Table 1 lists some local hospitals that have traditionally accepted student volunteers.

NOTE: *If possible, hospital volunteer work should continue through the summer and/or through at least two years of your college education.*

Table 1: Opportunities for Hospital Volunteer Work

Hospital	Phone Number
Abington Hospital	215.576.2490
Albert Einstein Medical Center	215.456.6058
Children's Hospital of Philadelphia	215.590.1093
Hospital of the University of Pennsylvania	215.662.2575
Lankenau Hospital	215.645.2138
Medical College of Pennsylvania	215.842.6425
Thomas Jefferson University Hospital	215.955.7777

Figure 10-5: Procedure for applying to medical school.

Sophomore Year

1. Take the last introductory biology course (Bio 230), a biology elective, and organic chemistry (Chm 201 and Chm 202). If you have not taken calculus (Mth 120) or computer science (Csc 152), take them now.

2. Continue your involvement in extracurricular activities and volunteer work.

3. Sign up for the college's mentoring program. This program will set you up with a physician who is an alum willing to answer any of your questions and show you what he/she does.

4. Obtain a summer research job. While research is not a requirement for getting into medical school, it is very beneficial both for your application and for your education.

Junior Year

1. Take two biology electives (preferably Bio 412 and Bio 418) and physics (Phy 105 and Phy 106).

2. Continue your involvement in extracurricular activities and volunteer work.

3. Sign up for an advisor from the Pre-Health Professions Advisory Committee.

    NOTE: *After you receive your advisor, you should set up regular meeting times with him/her to further discuss the application process and begin writing your personal statement.*

4. Sign up and attend one of the MCAT review courses offered on campus (Kaplan or Princeton Review). While taking a review course is not mandatory, it is very beneficial to students and has traditionally raised MCAT scores.

Figure 10-5: Procedure for applying to medical school (*continued*).

Components and Structure

Instructions share some common components that generally follow a chronological order. The basic components include an introduction, which gives an overview of the process and lists any needed equipment or tools. A series of steps arranged sequentially and preceded by notes, cautions, warnings, or dangers, if needed, usually follows the introduction. Instructions often conclude with either a brief summary of the steps, a description of the result, or product information on maintenance or repair.

Introduction

The introductory section can range in length from a short paragraph to several pages depending on its subject. If your readers need specific equipment, tools, or other prerequisites to complete the instructions, you can list them here. Figure 10-6 is an example of an introduction to instructions on using a stemming machine to prepare floral bouquets. In this example, the introduction gives an overview of the machine, defining what it is and explaining its purpose. Then a list of required tools or equipment follows (see Figure 10-7).

How to Use a Stemming Machine

Introduction

A stemming machine is a non-motorized machine used in the floral industry. It is a heavy precision-built machine with a cast iron body and four rubber feet to hold it in place. The machine is used to place a 1½ inch steel pick on a silk or dried flower stem to provide a secure insertion into a floral oasis (a dry foam situated in a container). Making a secure insertion into a floral oasis will insure the sturdy construction of a floral arrangement. The only action involved is gripping the attached handle (Figure 2) to apply a simple up and down motion to crimp the steel pick around a stem.

Note:
Packages of steel picks require careful handling. The steel picks are packaged in stacks of 100 and are very sharp. Wear lightweight work gloves when placing a stack of picks into the machine to avoid cutting your fingers.

Figure 10-6: Introduction to instructions on using a stemming machine.

Requirements

- Stemming machine
- Stack of steel picks
- Steel pick weight
- Silk or dried flower stem

Figure 10-7: Equipment required for using the stemming machine.

Steps

Steps or directions should appear in some sequential order, often with numbers to make their sequence clearer (see Figure 10-8). In devising these steps, you might want to first look at the result and then work backwards, identifying each action that led up to it. Whatever approach you use, keep in mind that the steps need to be discrete and not overlap.

If any of the steps are complex and involve more than one action, try dividing them into several smaller steps, and listing these sub-steps under the larger ones.

Steps

1. Place the stemming machine on a flat surface.

2. Slide stacked steel pins down the magazine and put the weight over the picks.

3. Rest the flower stem in the v-shaped slot in the front.

Figure 10-8: Beginning steps in using a stemming machine.

Notes, Warnings, Cautions, and Dangers

At the beginning of some steps and following the Introduction, you might need to add a **Note** or a statement labeled **Warning**, **Caution**, or **Danger**. While the meaning of these terms might seem arbitrary, distinctions exist among them. A **Note** is almost always used to indicate additional or helpful information. A **Caution** is used to denote possible harm to equipment, and a **Warning** is used to denote possible minor bodily injury.[1] **Danger** is usually reserved for a serious situation—an

---

1　"Summarization Re: Standards for graphic symbols, etc." TECHWR-L. 12 February 2005. http://www/techwr-l.com; "UIP200 User's Manual: Introduction–Document Conventions" Uniden America Corporation, May 2004. 12 March 2005. http://www.uniden.com/pdf/UIP200om.pdf.

imminently hazardous one, "which if not avoided, will result in death or serious injury."[2] Figure 10-9 contains an explanation of some of these terms that you might place at the beginning of a manual.

Remember always **place these terms prominently before the step or steps to which they apply**. You want to make sure that the person reading your instructions is aware of any harm that can occur *before* performing a particular step or series of steps. You want to not only act ethically and responsibly but also avoid having a lawsuit filed against you or your company because of any resulting harm.

Introduction

Document Conventions

Warnings, Cautions, Notes, and Timesavers

This document uses particular conventions for warnings, cautions, notes and timesavers. Below is an example of these conventions and how they are used:

> **WARNING**
> *Warnings appear with the stop sign. They indicate there is extreme danger or a risk of injury if the instructions are not followed correctly. Some warnings are required by government regulation or company policy.*

STOP

> **CAUTION**
> *Cautions appear with the "slow" sign. They indicate there is a danger of damaging the equipment or causing a mission-critical problem if the instructions are not followed correctly. Some cautions are required by government regulation or company policy.*

SLOW

NOTE
Notes point out important and useful information that the user should remember, but there is no danger. Most notes are based on by suggestions from users or frequent questions to support personnel.

TIMESAVER
Timesavers are hints that will make things go a little faster. Most timesavers are suggestions from users that have worked well in the field.

Figure 10-9: Document conventions for Uniden's UIP200 user's manual (http://www.uniden.com/pdf/UIP200om.pdf) *Reprinted courtesy of Uniden America. Inc.*

---

2   ANSI Warning Labels, Danger Labels, Caution Labels." Maverick Label.com. 12 February 2005. http://pfl.labelserve.com.

Conclusion

The concluding section might describe the end result of the process or include a "trouble-shooting" section or information on maintenance and care (see Figures 10-10 and 10-11). It can be a separate paragraph or just a sentence or two attached to the final step.

Conclusion

The operation of the stemming machine is simple and fast. It is the best device available to assist in arranging silk or dried flowers quickly.

Preventive Maintenance and Care of the Stemming Machine

Clean and oil the machine periodically. To clean the machine, use any standard grease remover or cleaning solvent. To oil the machine, simply oil all moving parts.

A thin film of "3 in 1 Oil" is recommended.

Figure 10-10: Conclusion of instructions for using the stemming machine with maintenance information.

After you've finished with your arrangement, don't forget to use a watering can or a tall glass to put some water and fresh flower food in the vase. The colder the water is the better; flowers stay fresher longer in chilly water because it keeps the buds closed.

Now that you've learned the basics of putting together a flower arrangement, you can try out different flowers, colors, shapes and decorations. There are no concrete "rules" to flower arranging, just some helpful guidelines. So, experiment with different flowers and see what *you* like.

Figure 10-11: Conclusion of instructions for making a floral arrangement with tips for keeping the flowers fresh.

Writing Text

Sequential Arrangement

Arrange your steps in a logical order so that each new step builds upon the result of the previous one. Generally **order them chronologically**, but if you can complete some steps in a different order, tell the reader. Use whatever order is easiest to follow.

Voice, Person, and Mood

Use **active voice, 2nd person "you,"** and **imperative mood (a command or polite request)**. For example, write *Push the lever to the right* rather than *The operator should push the lever to the right* or *The lever should be pushed to the right.*

One Major Action

Limit each step to one major action. Write *Push the lever to the right* – not *Push the lever to the right, and raise the bar to the next level.* However, if you can perform two actions simultaneously, use both actions in your sentence: *While pushing the lever to the right, raise the bar to the next level.* If you can divide the step into two or more sub-steps that clearly are part of a single overall instruction, give the overall step first. Then follow it with the sub-steps, possibly labeling them as "a" and "b" or distinguishing them in some other way to show their subordinate position in the sequence. For instance,

1. Restart the computer.
   a. Click onto the Start menu.
   b. Click "Turn off the computer."
   c. Click "Restart."

Parallel Structure

Make sure you use parallel structure in composing the text for your steps. **Begin with an action verb for every step, and use the same structure throughout.** If you follow a step with an explanation in a complete sentence, then follow all succeeding steps with explanations in complete sentences. If you follow a step with a phrase, use phrases after all steps.

Numbering

Use **Arabic numbers** for steps, and **letters** or **smaller-sized numbers** for any sub-steps that you might include. Numbers will make the process seem less intimidating, and will keep the reader aware of his or her progress in completing the procedure. In Figure 10-12, the large number "7" distinguishes the major step of installing software for a printer, while smaller numbers identify the substeps needed to complete this step.

Designing Pages

When laying out pages, consider the size of the page that the instructions will be printed on, and where they will be used—outside, inside, in well-lit areas, or in dark rooms. Also consider who will use the instructions—a very technical user or general consumer. These considerations will influence your choice of type font, size, and column width.

7 Install Your Windows Software

Locate the CD-ROM that came with your printer. Check to see which version of Windows you have, then follow the right set of instructions for your system.

Windows XP

Your computer may have the EPSON printer software already installed. **Make sure the printer is still turned on, then turn on your computer.**

If you see this message in the bottom right corner of your screen . . . .

You're ready to start printing!

If you see this screen . . . .

Click **Cancel**. Then follow the steps below to install printer software:

1   When you see the Windows icons on your monitor, put the EPSON CD-ROM in your CD-ROM or DVD drive.

2   If you see a screen asking **What do you want Windows to do?**, click **Cancel**.

3   Click **Start**, click 🖥 **My Computer**, double-click 🖴 **EPSON**, and then double-click 🖴 **EPSON**.

4   Read the license agreement and click **Agree**.

5   At the Main Menu, click **Install Printer Driver**. • • • • • • • • Wait while the files are copied to your computer.

6   If you see this screen, make sure the printer is connected and turned on. Wait for a minute for the screen to disappear. Do **not** click **Stop searching**. • • • • • • • • • • •

7   When you see a message that setup is complete or the printer port has been set, click **OK**. The EPSON Product Registration screen appears.

8   Register your printer as instructed. At the last screen, click **Done** or close your browser. Now you're ready to print!

9   To place a link to EPSON's free photo-sharing website on your desktop, click **Share Photos Online at EPSON PhotoCenter**. On the next screen, click **Add Icon**, then click **OK**.

10   Click **Back**, and then click **Exit** to close the Main Menu. Remove the CD-ROM. You're ready to start printing!

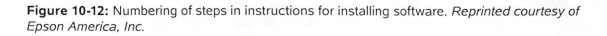

Figure 10-12: Numbering of steps in instructions for installing software. *Reprinted courtesy of Epson America, Inc.*

Using Color

Color, as explained in Chapter Six, can orient your reader. With instructions, color can work effectively when used to label parts that need to be assembled or to code major sections of your procedures. Figure 10-13 shows a screen with a procedure on formatting the different parts of bibliographic entries using MLA style. If you go to the actual website (http://www.liunet.edu/cwis/ cwp/library/workshop/citmla.htm) you will see each part appears in a different color.

Color is also important in distinguishing Warnings, Cautions, and Dangers. While organizations may differ in what they use these terms to convey, most agree on the use of certain colors to designate more serious situations.

The color **yellow**, whether it's used for "Caution" or "Warning," indicates the least hazardous situation, with **orange** designating a more serious one. The color **red** traditionally indicates the most hazardous situation. Figure 10-14 shows examples of introductory statements about possible harm from using a chemical paint remover, where "Caution," "Warning," and "Danger" would appear in these colors.

Figure 10-13: Instructions for MLA citations. *Reprinted courtesy of B. Davis Schwartz Memorial Library, Long Island University: http://www.liunet/edu/cwis/cwp/library/workshop/citmla.htm.*

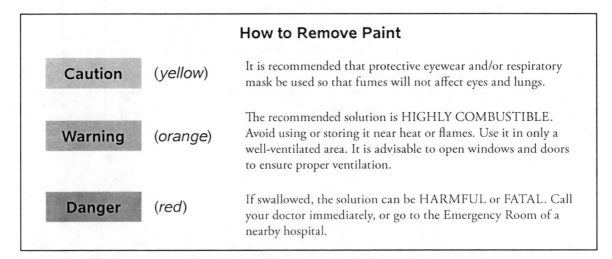

How to Remove Paint

Caution (*yellow*)	It is recommended that protective eyewear and/or respiratory mask be used so that fumes will not affect eyes and lungs.
Warning (*orange*)	The recommended solution is HIGHLY COMBUSTIBLE. Avoid using or storing it near heat or flames. Use it in only a well-ventilated area. It is advisable to open windows and doors to ensure proper ventilation.
Danger (*red*)	If swallowed, the solution can be HARMFUL or FATAL. Call your doctor immediately, or go to the Emergency Room of a nearby hospital.

Figure 10-14: Warning, Caution, and Danger given at the beginning of instructions.

Checklist

	Do You Know	Yes	No
1.	the different types of instructions?		
2.	what should be included in an introduction?		
3.	what is important about listing the required parts or equipment?		
4.	why the steps of the process need to be logically and sequentially arranged?		
5.	what purposes a conclusion can serve for a set of instructions?		
6.	the difference between notes, warnings, and cautions?		
7.	why notes, warnings, and cautions need to be placed before steps and prominently displayed?		
8.	why steps need to be written as commands (2nd person, active voice)?		
9.	why maintaining parallel structure is important?		
10.	why you want one main action verb for each step?		
11	why main steps should be numbered, and how you can identify the sub-steps?		
12.	why you want to consider where the instructions will be used when designing pages?		
13.	why graphics need to be placed near the text?		
14.	how color can work to distinguish sections of the instructions?		
15.	what colors indicate degrees of possible harm?		

Exercises

1. Bring to class a set of instructions you've used in the past for assembling, installing, or operating a product. Are they easy to follow or difficult? What do you think accounts for your response—the text, graphics, or both?

2. Do introductions to instructions need to be long? If you have limited space, what do you think is most important to include?

3. Give examples of occasions when you might use a Note, Caution, Warning, or Danger with instructions, and explain why colors used for each are important.

4. Go to the Website constructed by the Department of Energy for the Yucca Mountains Nuclear Waste Repository (http://www.ocrwm.doe.gov/ym_repository/index.shtml), and look at the color scheme used. What effect do the earth tones, blues, and pinks have when reading about the safety of handling nuclear waste?

5. How might you create instructions to be used by people who speak different languages? Would you translate each step into the different languages of your audience? Would you eliminate words and use only graphics? Discuss possible ways of creating instructions that are cost effective and yet clear to readers in different countries.

6. Working in groups, evaluate several instructions for products you've purchased for your home.
   - Consider the graphics, layout and design, the use of color, and arrangement of steps.
   - Identify instructions you think are attractive, clear, and easy to follow and those you find uninviting and difficult to follow. Explain why.

7. Evaluate the graphic in the instructions in Figure 10-15. How effective is it in explaining the second step?

8. In Figure 10-15, evaluate the writing. Is it clear? Could it be improved?

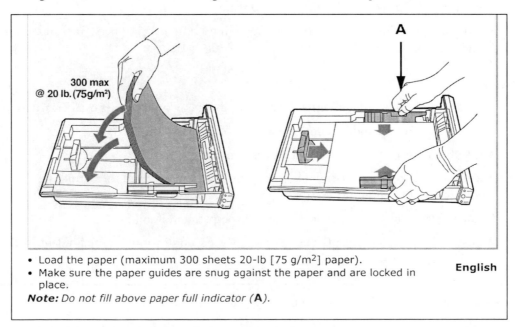

- Load the paper (maximum 300 sheets 20-lb [75 g/m²] paper).
- Make sure the paper guides are snug against the paper and are locked in place.

Note: *Do not fill above paper full indicator (**A**).*

English

Figure 10-15: Loading paper into a printer's tray. *Reprinted courtesy of Oki Data Americas, Inc.*

9. Prepare a set of instructions explaining something that you know well. These should be written for a general audience whom you tell how to assemble, install, or operate something or how to perform a procedure.

   - Be sure that the steps are arranged logically.

   - Use visuals, and place them close to the text that they illustrate. If possible, create your own. However, if you do download them from the Internet or take them from another source, be sure to credit the source.

   - For the text, use active voice, 2nd person *you*, and parallel structure.

   - Use notes, cautions, warnings, and dangers, if necessary.

   - Provide sufficient background information and adequate details and examples so that your procedure is clear to the general reader.

10. Review the procedure below for a college student to respond to a written message containing a threat (Figure 10-16). How do these instructions follow the guidelines explained in this chapter? How might they be improved?

PROCEDURE

Printed threatening messages

1. Call Campus Police, and if you are a dorm resident call your R.A.

2. If possible a copy of the letter or picture of the message (if graffiti) should be sent to Student Services.

3. If graffiti, remove it or call Maintenance to remove it.

4. A written report should be filed by the responding Campus Police personnel and sent to Student Services.

Actions

1. Keep any proof or record of witness testimony about any prior threats or altercations.

2. Call Campus Police, and, if a dorm resident, notify the R.A.

3. The Office of Student Services should be notified by the R.A.

4. The head of Campus Police should meet the student who received threat and the Coordinator of Community Affairs.

5. Campus Police should take action to find the perpetrator, if unknown.

6. Action will be taken against the perpetrator by the Coordinator and the Student Review Board, which will get involved.

Figure 10-16: Procedure for students to report threatening messages.

CHAPTER ELEVEN

Proposals

Proposals come in many forms and have many applications. However, there are some basic elements that all share. This chapter focuses on different types of proposals, the steps involved in creating them, and the components that comprise them.

Objectives

- ✓ Learn the basic types of proposals and consider the needs of audiences for each of them.
- ✓ Understand the four basic stages involved in creating proposals.
- ✓ Learn the basic components of proposals.
- ✓ Design pages and integrate graphics within them.

Types

You can classify proposals according to whether they are solicited or unsolicited, sent internally or externally, or created to gain approval for research or for sales and funding.

Solicited Versus Unsolicited

Often you'll send a proposal to someone to gain a sale or get approval for implementing some project without being requested to do so. This **unsolicited** proposal is quite common when you see a need that you can fulfill.

However, more often you'll send out proposals to organizations that request them. These **solicited** proposals are in response to a document called an *RFP (Request for Proposal)* that corporate and non-profit organizations or government agencies issue. (Figure 11-1 provides an example of a RFP from Oxmoor House.) RFPs often share the following characteristics:

- They are usually detailed and lengthy.
- They ask for specific information to be supplied in a particular order on all aspects of the product or service proposed.
- They usually have a deadline for submission.

When you submit your proposal, the group that sends out the RFP examines and compares each section of it with those submitted by others.

Oxmoor House
List Management Request for Proposal

Introduction and Intent of Proposal

Oxmoor House is conducting a review of their list management business. Lists and segments currently available for rental are detailed on the enclosed data cards. Oxmoor House products are sold via direct mail, print advertising, outbound telemarketing, and Internet marketing. Names are available for rental for direct mail only. Names are not available for telemarketing or e-mail offers. Oxmoor House must approve all list orders.

Product lines include pre-notification negative option book series and continuity programs and a growing one-shot segment.

The process for review is detailed later. We would like to start with a written response to this RFP. That will be followed by possible presentations here in Birmingham and visits to your offices. We will be going through this process with several List Managers and at each step in the process we plan to narrow the field of candidates.

Our goal is to review the business with List Managers to identify opportunities to increase overall list rental revenue and to improve our representation in the direct marketing community.

Assumptions for Financial Projections

Please use the following assumptions when projecting annual list rental revenue.

- Approximately 23% of the business is on exchange. If overall rental business grows, we will expect this percentage to decline. Less than 5% of the business is to other Time Warner entities and is at sister company rates. If overall rental business grows, we also will expect this percentage to decline.

- List files are mostly U.S. addresses. Our goal is to grow our Canadian business in the next 2-3 years.

- The commission structure should be negotiable and must be addressed in the response.

- If modeling or other techniques are to be assumed in the financial projections please isolate the value and costs, e.g. cost of models, data overlays. We currently offer regression-modeling services. Current pricing: we build models for free and there is a $10/M regression select charge. Oxmoor House staff does all modeling.

- Oxmoor House does all list rental fulfillment. Our list rental team is based in Birmingham.

Oxmoor House Books List Management RFP -
Confidential & Proprietary   **1**   09/26/02

Figure 11-1: First page of a corporate RFP. *Reprinted courtesy of Oxmoor House.*

Internal Versus External

Another classification of proposals depends on the location of the person(s) to whom you send them. A proposal that you send to your supervisor, colleagues, or another department within your organization is an ***internal*** one (see Figure 11-2). On the other hand, an **external** proposal is one you

To: Department of English

CC:

From: Mary Thompson

Date: October 16, 200X

Re: PROPOSAL FOR A SPECIAL TOPICS COURSE

Of the 100 recipients of the Nobel Prize for Literature, ten have been women. These women come from nine different countries on four continents and span the period from 1909 to 2004. As writers, what do they have in common? What themes do they convey in their works, and what literary characteristics and concerns do they share with their contemporaries? To answer these questions, I propose a course entitled "Women Literary Laureates."

This course will attempt to answer these and other questions by looking at the works of these ten literary laureates and focusing more in depth on the fiction and poetry of five of them. Students in this course will study the major fiction of Pearl Buck (U.S.A., 1938), Nadine Gordimer (South Africa, 1991), and Toni Morrison (U.S.A., 1993), as well as the poetry in translation of Gabriela Mistral (Chile, 1945) and Wislawa Szymborska (Poland, 1996). Additionally, students will select one of the works of the other five laureates (Selma Lagerolf, Grazia Deledda, Sigrid Undset, Nelly Sachs, and Ellfriede Jelinet) and research and report to the class on their reading. Throughout the course, the objective will be to look at these authors' works for their literary merits and to evaluate aspects that may have led to their authors' being selected for the prestigious Nobel Prize.

Tentative List of Readings
Required works
Pearl Buck, *The Good Earth***
Gabriela Mistral, *Selected Prose and Prose Poems*
Nadine Gordimer, *Selected Stories* * and *The Conservationist*
Toni Morrison, *Beloved***
Wislawa Szymborska, *View from a Grain of Sand: Selected Poems*

Additional selections
Selma Ottilia Lovisa Lagerolf, *The Wonderful Adventures of Nils*
Grazia Deledda, *After the Divorce*
Sigrid Undset, *The Bridal Wreath* (first volume of the trilogy *Kristin Lavransdatter*)
Nelly Sachs, *O the Chimneys: Selected Poems including the verse play "Eli"*
Ellfriede Jelinet, *The Piano Teacher***

*Since this edition is currently out of print, several stories will be duplicated and distributed.
**Films based on these novels will be viewed in the Connor Library.

Requirements for the Course
Exams
Quizzes (one on each of the five authors' works)
Final Exam
Papers/Reports
Short paper (3-5 pages) on the fiction or poetry studied in class.

This course meets the criterion for being "special" because these writers have not been studied before as a group. Although Szymborska may have been studied in English 338 and Morrison or Gordimer may have been covered in English 250, they have not been studied with the other women Nobel laureates.

Figure 11-2: First page of an internal proposal seeking approval for a new course.

send to a client or outside organization. While many of the features are the same in both types of proposals, obviously in an internal proposal, unlike an external one, you would not need to include information about your particular organization or about the qualifications of those who would complete the project.

Research Versus Sales

Another way you can classify proposals is by what you hope to achieve by having them accepted. Two of the most common types are those that seek approval for a **research** project and those that seek to gain a sale. While both need to be persuasive in tone, the **sales** proposal is usually more openly persuasive. Unlike the research proposal, a sales proposal focuses on costs and services or products provided rather than methodology. Figures 11-3 and 11-4 show segments of both research and sales proposals.

Formal Versus Informal

The examples in Figures 11-3 and 11-4 are **formal** proposals. However, proposals can also be **informal**, as are those prepared by vendors and service providers for potential customers of their products. As you can see from Figure 11-5, an informal proposal generally consists merely of a listing of the services to be provided along with their overall cost.

Creating Proposals

Nick Bernardo, a former proposal writer for NCO Group in Fort Washington, Pennsylvania, says that the process of preparing a proposal consists of four stages (see Figure 11-6).

Research

The research stage, according to Bernardo, takes about 25% of the time needed to complete the proposal and involves **talking to experts in the subject matter** within the company and with external consultants. Bernardo says it involves **gathering statistical data**, which consists of industry-specific information. He explains that this research comes from printed and online sources—periodicals and company Web sites, for example.

Writing

This part of the project, according to Bernardo, takes the most time (45%). He believes that the key to success in this part is to use standard copy for all company proposals and then customize it according to different clients and multiple audiences. Bernardo points out that in addition to the prospective client, these audiences might include the writer's boss and sometimes the boss's boss. Finally, Bernardo says that in writing a proposal, you want to distinguish carefully between facts and beliefs—what he terms the "semantics" part of the writing.

Proposal for Research Leave

Summary of Proposal

This proposal is to write a textbook tentatively entitled *Professional and Technical Communication: A Comprehensive Guide (see Appendix A)*. This textbook is designed for English or a Communication course for upper-division undergraduates or first-year graduate students in a professional or technical writing course. The approach used in this book encourages students to view both text and graphics or layout as one--as an integral part of the overall document rather than as separate entities with the emphasis placed on writing and with graphics viewed as merely add-ons or with the entire layout ignored. This book will cover material used for both business and technical writing courses as well as material for preparing promotional pieces--technical brochures and data sheets.

I have considered this project for sometime, especially because I have experience working in industry as a technical editor and publications manager as well as experience teaching technical and business writing. I did do preliminary research for a similar project in the summer of 1999 when I was given a grant that helped me to both prepare for this book and develop my courses in professional writing. Prior to the grant, I had taught English 303: Business Writing and English 409: Technical Publishing for only one semester. However, with the help of that grant I was able to update these courses and consequently have taught them in spring semesters for the last three years. With this experience and more recently with teaching Communication 603 (Strategies in Professional Writing), I feel qualified to take on this project.

I plan to begin writing this book next summer and seek a research leave to continue the writing throughout the fall of next year.

Semester for Leave

I would like to use the fall semester to pursue this project. I plan to begin the writing during the summer, and, if a research leave is granted, I can continue through the fall and return in the spring to teaching before preparing materials for review in the fall of the following year.

Substance of Proposed Study

The purpose of the study is to complete research and write several chapters of a textbook for professional and technical writing classes.

Research Accomplished

So far I have completed the following:

- a review of existing technical communication textbooks,
- research of sales of current textbooks on Amazon's and Barnes and Noble's Web sites,
- preparation of a prospectus and an annotated table of contents,
- gathering materials for chapters:
  - articles on the repercussions at the *New York Times* because of fabricated articles (Chapter 2)
  - student papers as examples for Chapters 10-15
  - survey of employers who recruit at Larson regarding the skills and experience they are looking for on résumés (Chapters 14 and 16)
  - letters of inquiry to major publishers of textbooks: Allyn & Bacon/Longman, Bedford St. Martin's, Houghton Mifflin, McGraw-Hill, Norton, and Prentice Hall.

Figure 11-3: First page of a research proposal.

Oxmoor House

Executive Summary

Unmatched Experience With Your Files

MKTG Services is a leading relationship marketing company with unsurpassed expertise in the publishing industry. Having helped build a number of your list management properties over the past eight years, MKTG Services has a firm grasp on the **Oxmoor House** files and the markets they serve. This depth of experience, along with our vast list management expertise, uniquely positions MKTG Services to market, advertise and sell the **Oxmoor House** customer lists and deliver the highest potential revenue possible.

Delivering Revenue Through Multiple Streams

Our list managers take pride in making highly accurate projections for our clients, based on detailed research and analysis. We have profiled your book buyers by available selection criteria, and based our projections upon our experience with your lists and our knowledge of the mailers most likely to use your files. Our revenue projections for the **Oxmoor House** lists have been enclosed for your review. Based on these projections, MKTG Services will allocate a portion of our anticipated commission toward the marketing and advertising of your files—giving us added incentive to meet our revenue expectations.

Proactive Marketing Across Multiple Channels

MKTG Services aggressively pursues sales through a comprehensive marketing strategy across multiple channels. We raise awareness of your lists through unrivaled convergent marketing, including premium-based direct mail promotions, telesales, space advertisements, email promotions, trade show representation and face-to-face sales.

Strategies That Make the Difference

The ultimate test of your list management program comes at the bottom-line. Our strategies ensure:

- Maximum list rental income
- Revenue-winning pricing, promotion, selling, and negotiation
- Complete penetration of primary markets
- Diversification into new markets
- Growth stimulation through aggressive testing

MKTG Services
www.mktgservices.com

MKTG SERVICES CONTACT: Denise I. Hubbard
PHONE: 215.968.5020 x114
EMAIL: dhubbard@mktgservices.com

3

Figure 11-4: First page of a sales proposal. *Reprinted courtesy of Oxmoor House.*

D & D FAMILY ROOFING

845 Willow Drive

South Hills, PA 19653

216-874-2354

SHINGLE SLOPE ROOFING

GAF Timberline 30 year warranty	$6000
Ventilation system	$ 524
Flat roof over garage (recoated)	$ 300
Total:	$6824

Price includes the removal of old roof, installation of new shingles, all labor, materials, and delivery.

Local permit and replacement of wood are not included.

Figure 11-5: Informal proposal from a roofing contractor.

Review

Depending on the number of reviewers and the number of times that the proposal is looked at, this phase can occupy 20% of the project. Bernardo points out that you want to ask before the review starts not only who is in the review loop, but also who will be looking at what parts of the proposal. In this phase, reviewers examine the document for factual content, grammar and punctuation, message or theme, and format. Bernardo cautions further that the writer must be prepared to accept constructive criticism.

Production

This last part takes about 11% of the process and includes duplicating and binding, and—in some cases—distribution. Bernardo warns that this last stage can sometimes take twice as long as you think it will and that you need to be prepared for such unexpected situations as scanners or copiers malfunctioning.

Putting Parts Together

Different parts exist in different types of proposals and within different organizations, but some similar parts appear in many of them. These parts are not necessarily those you'll need to include when replying to an RFP. However, if you're creating a proposal, particularly a research proposal, without any prescribed format, consider including some of the following.

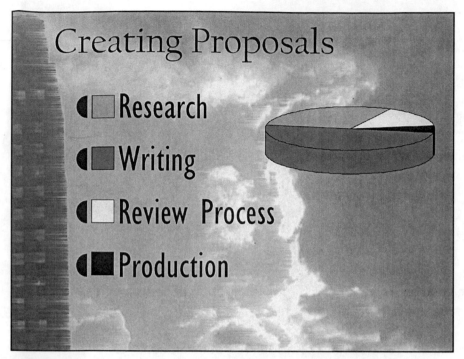

Figure 11-6: Nick Bernardo's four stages of creating a proposal.

Description of the Problem

A proposal often begins with a section describing a problem. Sometimes this problem appears in a background section, but often it appears as a brief overview.

Objective or Purpose

A statement of your objective often follows your description of the problem. It tells the reader what you hope to accomplish with the proposal.

Background

A summary of the background explains what led up to the proposal—why you're sending it. Background gives you the opportunity to elaborate on the history of the project—what has been done previously, what worked, and what did not work.

Figure 11-7 shows examples of the Background, Problem and Purpose sections that appear in a student's proposal for a feasibility project on expanding a peer-tutoring system.

Qualifications

Proposals often include some information on the qualifications of the people who are submitting them or who will perform the services or supply the products. In sales proposals, you might include an organizational chart that shows the relationships among managers within a company and gives the prospective customer a sense of how the company operates. Also, you might include a listing of the educational degrees of individuals who will be working on the project along with their experience and familiarity with the work. This section can appear either before the "Plan" and "Schedule" or after them.

PROBLEM AND OBJECTIVE

The Writing Fellows Program provides training and guidance for undergraduate peer tutors. These peer tutors read the drafts of students in one class, edit the papers for content and grammar, and meet with the students to discuss the alterations.

The Program, in my opinion, is beneficial to both students and professors. Students have access to extra help in revising their papers before handing them in to be graded; professors can be certain that their students have revised their papers at least once, and that the students are not handing in hastily written first drafts. The peer tutors benefit from the experience of revising others' writing as well. However, despite these benefits, the number of students interested in becoming Writing Fellows seems to be very few, and Dr. Severn has stated that she has had difficulty recruiting faculty sponsors.

My objective in this study is to analyze the feasibility of different methods of expanding the Writing Fellows Program to include both more student tutors and more professor sponsors.

BACKGROUND

The Writing Fellows Program is run by Dr. Margie Severn. This Program recruits and trains undergraduate students as peer tutors.

The Program began in the mid-1980s during a surge in the popularity of Writing Across the Curriculum programs. Professors from the English Department went to regular workshops at Brown University to learn more about Writing Across the Curriculum, a movement that strove to emphasize the importance of writing in all subjects, not just traditional subjects. Writing Fellows programs developed from the popularity of the Writing Across the Curriculum movement.

Students who are interested in becoming Writing Fellows must submit an application and two writing samples to Dr. Severn in the spring (usually by late February). Dr. Severn reviews the applications and accepts students into the Program based on their proficiency as writers.

All of the students who are accepted into the Writing Fellows Program must take ENG/ HON 360 - Writing and the University. This course, taught by Dr. Severn, trains students in different aspects of peer tutoring. Dr. Severn assigns Writing Fellows to faculty sponsors. The students of the faculty sponsors submit drafts, which the Writing Fellows read and edit for content and grammar, and the students meet by appointment with the Writing Fellows to discuss the revisions. Typically, the Writing Fellow reads 12-15 drafts and holds subsequent meetings twice a semester. Writing Fellows are expected to work approximately 60 hours per semester and are given a $300 stipend.

Dr. Severn theorized that the gradual decline in the use of the Writing Fellows Program has little or nothing to do with the quality of the work of Writing Fellows. She attributes the decline to two separate factors.

Figure 11-7: Problem, Objective, and Background sections in a student's proposal to research ways of expanding a peer-tutoring program.

Dr. Severn attributes the decline in the number of professors who utilize the Writing Fellows Program not to the quality of the work of the Writing Fellows, but to a change in the dynamics of the University's professors. When the Writing Fellows Program first began at Larson University in the mid-1980s, two-thirds of the faculty went to workshops to learn about Writing Across the Curriculum and how to incorporate the Writing Fellows Program into their classes. All of the new faculty hired went to these workshops. Since then, however, many of the faculty members that were very involved with the Writing Fellows Program have either retired or have been promoted and teach upper-division classes that are not conducive to the guidelines of the Writing Fellows Programs. For example, Dr. Morgan in the Philosophy Department taught 100-level core Philosophy classes and was an avid user of Writing Fellows. Presently, however, he teaches only upper-division philosophy classes and does not have a need for Writing Fellows in his classes. Writing Fellows traditionally are supposed to read two papers of about three to four pages in length in a semester. Often, upper-division classes have much longer and more complex papers.

Since the surge of interest in Writing Across the Curriculum has declined somewhat, interest in the Writing Fellows Program has declined as well. Faculty members who have been hired more recently have not had the opportunity to attend the numerous workshops that were available in the past. Because of this, professors may not be sufficiently aware of the Writing Fellows Program and its benefits.

The decline in the number of students who apply to and participate in the Writing Fellows Program can be attributed to the increasing number of activities available to students on campus, changes in students' interest in certain majors, and more rigorous curriculums in the more popular majors.

The activities that students can participate in on campus have grown far more numerous in recent years. The University's student newspaper, *The Collegian*, the student theater group, the Masque, along with the Resident Student Association and the Ambassadors are all popular activities that place significant demands on a student's time.

In addition, there have been significant changes in the number of students that major in certain subjects. The number of students majoring in English, the most common major among Writing Fellows, has declined. Nursing and Education are two of the most popular majors. The rigorous coursework in these majors leaves students with less free time. Because Nursing and Education majors have more required classes than other majors, there are fewer free elective slots to take ENG/HON 360, the class required to become a Writing Fellow. These factors have led to a gradual decline in the number of students and faculty sponsors taking part in the Writing Fellows Program.

Figure 11-7: Problem, Objective, and Background sections in a student's proposal to research ways of expanding a peer-tutoring program (*continued*).

Data Sources or Methodology

In a research proposal particularly, you want to give the sources for the information that you gather. You can include both *primary* and *secondary sources* or mainly one or the other. (See Chapter Three for a definition of these terms.)

Scope

If you include "Scope," you're describing what is covered in the proposed offer and what is not covered as well as the reasons for including certain items and excluding others. Often this section is called "Scope and Limitations" or "Limitations" because it explains the extent of your study—in other words, what limits you impose on the research or product and service.

Figure 11-8 provides a part of a student proposal wherein "Scope" follows "Qualifications" and "Data Sources."

Plan

The "meat" of the proposal is the plan itself because it outlines the steps that you'll take, services you'll provide, or specific details of the product that you offer. In this section, you have the opportunity to describe fully what you want to do or sell in such a way to impress your reader. Figure 11-9 shows a plan for a feasibility study on the effects of medication used to treat Attention Deficit Disorder that is impressive with its organization and details.

Schedule

The steps you'll take are often paired directly with the timeline for their completion. If you give the schedule separately, however, you might put it in a list or chart. Figure 11-10 shows such a chart—a timetable a student created for completing her feasibility study during Spring Semester.

Writing Text

Using Tense

Use present and future tenses for writing most proposals. Obviously, when you write the "Background" section, you'll need to use past tense. However, when you describe your plan itself, you want to write in the present; and when you discuss expected results, you use future tense. When shifting from one tense to another, avoid awkward shifts that result from changes within a sentence or midway through a paragraph.

Checking Accuracy

While all your documents need accurate details, proposals especially need them because they function as legal documents. If what you offer is incorrect or you're unable to fulfill some of the conditions in a proposal after it's accepted, you not only have acted unethically but have become liable for legal action being taken against you or your company. Remember that a **proposal is considered a contract**, so check out any parts that appear questionable or which you might be unable to fulfill.

QUALIFICATIONS

Although I am far from an expert in photography, I did take photography courses both in middle school and high school. At both of these schools, students were provided with all the tools and equipment necessary to learn about darkrooms and photo developing. Additionally, I grew up with two parents who are artists. Over the years, they have dabbled in photography, and I have learned bits and pieces from their endeavors. Above all, my love for art and appreciation for the traditional methods of film developing has given me the motivation and determination to investigate this subject, in hope that I can make an educational change for students of the future.

DATA SOURCES

Interviews

Interviews will be conducted with several individuals on this campus. I will be working in conjunction with Dr. Habler, the Director of the Fine Arts Program, as well as other individuals in the department. I hope to interview members of the funding board and Dr. Joseph Connors to seek advice and direction about budget information and student interest. Additionally, I will conduct an interview with three experts in darkroom construction in order to compile a list of necessary equipment and procedures when building a photography lab.

Emails

I will be in contact with several people via email. I have already emailed the Art Department at the following universities to see the status of their photography courses and labs: Drexel, St. Joseph's, Temple, Philadelphia, and Villanova. I will be corresponding with retailers of photography equipment and construction experts in order to compile an estimate for the entire project. I will also be emailing Dr. Connors in regards to seeking approval for the room location and reconstruction; Dr. Connors will also be asked what alternatives exist if a room cannot be constructed or refurbished.

Phone Calls

Phone calls will be made, as needed, to several construction companies as well as to several of the above named universities.

Internet

One of the most important parts of this project will be to research and read about darkrooms and what their construction entails. I will scrutinize the available information to determine the best possible plan. I will be comparing prices and layouts for darkroom construction in order to develop a range of options for our school.

SCOPE

This study is limited to building a room or converting an existing room on campus. There has been some discussion about Larson purchasing Gemini Hospital, but I am unsure of this site as a potential location at this time. The study is not to examine the possibility of constructing a darkroom off-campus. The studio art building located behind the Harwood Center has been considered as a possible site for the construction of the lab; however, the building may not be adequate because of the plumbing problems and poor condition of its basement. Thus, it is not included in this study.

Figure 11-8: Qualifications, Data, and Scope sections in a student's proposal to study the feasibility of creating a photography lab on campus.

PLAN

I plan to research this topic in several ways:

- I will conduct a two-week study beginning on March 3 of a child with Attention Deficit Disorder monitoring his hours of sleep, calorie intake, and behavior on and off the medication.
- I plan to speak to a child psychiatrist and social worker who both specialize in working with these children.
- I intend to interview and/or survey parents and teachers of children with Attention Deficit Disorder.
- I will consult secondary sources that have been written on this topic.
- I will analyze the information I have obtained and present my draft and final copy of the report on this study.

Conduct a two-week study

For the two-week period from March 3 to March 17, the child will be observed for the following items:

- hours of sleep per night
- caloric intake per day
- description of behavior for the day

These factors will be monitored for 7 random days on medication and 7 random days off medication. Any unusual side effect of the medication or unusual behavior will be accounted for under "behavior," while the two major side effects of lack of sleep and low caloric intake will be evaluated separately. The child will be my brother, who is an 11-year-old with Attention Deficit Disorder.

Interview experts

Dr. Hawes, a child psychiatrist employed at Children's Seashore House in Atlantic City, has been my brother's physician for five years. Dr. Hawes treats many children with ADD. She also has been chosen to speak at many conferences concerning Attention Deficit Disorder. Dave Smith is a social worker who specializes in children with ADD.

Smith also works out of Children's Seashore House in Atlantic City. His main focus is on talking to the children to see how content they are at home, at school, and with friends or activities outside of school. I will interview both of these professionals, who can provide valuable information on the medical and social adjustments the child undergoes with and without medication.

Survey parents and teachers of children with ADD

I will interview and/or survey several parents and teachers of ADD children. Parents and teachers are vital sources of information concerning the behavior of ADD children because they are with the children for extended periods of time. Both parents and teachers are normally involved with the physician in determining the type and amount of medication

Figure 11-9: The Plan in a student's proposal for conducting a feasibility study on the effects of medication on a child with Attention Deficit Disorder.

which is given to the child. Teachers play an important role because they are able to witness changes in behavior and the concentration of the child in a classroom setting.

Examine secondary sources

With the increased number of children diagnosed with ADD, there has been an increase in the literature written about the disorder. Much of this information pertains to the pros and cons of medication and the effect that the disorder has on the child's performance in school.

Some sources that will be used in this study are *Maybe You Know my Kid* by Mary Cahill Fowler and *ADD: Helping your Child* by Warren Umansky, Ph.D. and Barbara Steinberg Smalley. Also two articles from *Pediatrics Review* will be consulted: "Children Who Have Attentional Disorders: Diagnosis and Evaluation" and "Children who Have Disorders: Interventions." These books and articles represent up-to-date information concerning the advantages and disadvantages of medication.

Prepare a draft of the feasibility report

The data obtained from the two-week period and the information from the other sources will be used to write my recommendation on how and if medication should be used for children with ADD. I will base my decision upon the criteria of hours of sleep and caloric intake versus behavioral changes. From the two-week study, I will have obtained data from my parents and teachers, which will aid me in my decision-making process. Alternatives to medication will be based on secondary research and the opinion of experts I have spoken with. All of this information will be compiled and used to write my rough draft, which will be submitted on April 1.

Confer with my professor and prepare the final draft

I will meet with Dr. Jones the week of April 8 to discuss my draft and possible changes to improve it. After my conference, I will begin revisions and add any additional findings from my research. I will then write my final draft of the report, which is due on April 15.

Figure 11-9: The Plan in a student's proposal for conducting a feasibility study on the effects of medication on a child with Attention Deficit Disorder (*continued*).

Choosing a Point of View

While a very formal proposal requires that you write almost totally in third person (*he, she, it, they, him, her,* or *them*), with a less formal proposal you might write in a mix of third and first person (*I, me, we, he, she, it,* or *they*). Certainly, if you're writing a proposal for a research project to your advisor or instructor, you would use "I," as you would in an e-mail or letter. In describing the problem and background about a topic you know first-hand, you would obviously need to use "I."

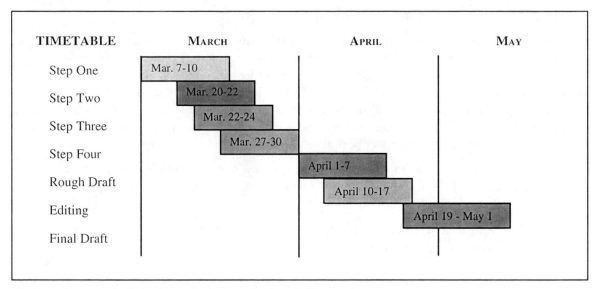

Figure 11-10: Schedule for completion of a feasibility study.

Designing Pages and Using Graphics

Designing Pages

To create a visually pleasing proposal, you want to integrate graphics and graphic elements throughout it. Consider laying out pages with your charts and other graphics carefully placed in ways that break up solid blocks of text. As suggested earlier in Chapter Four, you might plan your pages using a storyboard approach with text and graphics carefully positioned to complement each other.

Selecting Types of Graphics and Graphic Elements

Photos seldom appear in a proposal; and if you use them, you want to use them sparingly. More likely, you'll have charts or tables, and sometimes drawings. However, to make your proposal attractive as well as readable, you want to use graphic elements (rules, bullets, and different typefont sizes) throughout.

Tables and Charts

Often proposals include tables—both informal listings and more formal ones— that can give detailed information about your schedule, the costs, or any other numerical information you have.

You might include organizational charts that show the relationships of departments and managers to each other. Also you might include line and bar charts showing changes over time that appear in a Background section or even within the Plan to show projected returns. In addition, you most likely will give the schedule in the form of a timetable or other type of chart as shown earlier in Figure 11-10.

Rules, Bullets, and Typefonts for Headers

While text dominates most proposals, graphic elements are still an integral part of them. **Rules**, for example, at the top and bottom, between sections, or on the sides of pages serve not only to separate information, but to lead the eye of the reader into your document. Listings clarify information and generate additional white space. **Bullets** help distinguish items in these listings.

Varying sizes of **type fonts** and shadings help differentiate levels of your **headers**. Headers serve to orient the reader to the outline that underlies your content, and provide contrast with body text.

When creating headers, consider both type and color:

- Use a type family with several members so that you can differentiate the levels of headers with these different typefonts: for example, Franklin Gothic Book, Franklin Gothic Medium, or Franklin Gothic Heavy.

- Use color to contrast the black type on the page. Just two colors—black and blue, for example—can be quite effective.

Checklist

	Do You Know	Yes	No
1.	the difference between the solicited vs. unsolicited, internal vs. external, and research vs. sales proposals?		
2.	four stages for creating a proposal?		
3.	how much time should be given to research for a proposal?		
4.	what different audiences you need to consider while writing a proposal?		
5.	what is examined during the review cycle?		
6.	what production involves?		
7.	what parts need to be stated in the proposal's Introduction?		
8.	when to include a Qualifications section?		
9.	what type of proposal would include a Methodology section?		
10.	why the Plan and Schedule are the main parts of a proposal?		

	Do You Know	Yes	No
11	what tense to use for proposals?		
12.	why a proposal is considered a legal document?		
13.	what kind of graphics appears in proposals?		
14.	why headers help reveal the proposal's structure?		
15.	what purpose color serves?		

Exercises

1. Give several instances where different types of proposals (solicited or unsolicited, internal or external, and formal or informal) can overlap.

2. In a small group, discuss which type of proposal would be most appropriate for the following:
    a. the purchase of new copier for your department
    b. the remodeling of an office
    c. the feasibility of building a new stadium

3. Review the four stages of creating a proposal. Which do you think is the most important, and do you agree with the percentages that Bernardo assigns to each?

4. Why are the Plan and Schedule the most important parts of a proposal?

5. Why would you include Qualifications in a proposal?

6. Why might you be sued if you were unable to deliver products or services you listed in a proposal at a specified cost? What might you have done in preparing the proposal to prevent this legal action?

7. What type of graphics do you think are most common in proposals?

8. Prepare a one-page proposal for future research in one of your courses. What information would you include? What information would you not include?

9. Prepare a proposal (approximately 2-5 pages) to gain approval from your instructor in this course for your final project. Include all the components necessary for such a research proposal:

    - Statement of the Problem and Purpose
    - Background
    - Qualifications
    - Data Sources or Methods
    - Scope and Limitations
    - Plan
    - Schedule

10. Create several graphics to be included in this proposal.

CHAPTER TWELVE

Reports

At one time or another you'll need to prepare a report. These reports can range from a one-page summary of a trip to a complex analysis of a client's needs comprising several hundred pages. The proposal, discussed in the previous chapter, is one type of report; however, this chapter focuses on additional types, particularly the short progress report and the longer feasibility one.

Objectives

✓ Identify the different types of reports and learn ways to organize them.

✓ Prepare short reports such as those summarizing a trip, activities, or progress on a project.

✓ Prepare a longer formal analytic report with its preparatory and supplementary sections.

✓ Understand how to integrate graphics and graphic elements with text within short and long reports.

Types

Informal Versus Formal

You might report on individual activities or transactions informally. If so, you'll probably limit them to a few paragraphs or a page and just summarize what you've done. On the other hand, for a more intricate project, you might produce a formal report with headers, graphics, and a fixed organization.

Informational Versus Analytical

When you create a report, you provide information—information on a research project, a company's accounts, services completed, or the yearly revenue and expenditures sent annually to stockholders. However, sometimes when you create a report you provide an analysis of information collected. In these reports you interpret the information and draw conclusions that lead to recommendations. In these analytical reports you can give analyses of problems and offer solutions, compare several alternatives and determine which is best, or determine the feasibility of a proposed project.

Long Versus Short

Formal reports, especially those that analyze as well as provide information, are usually longer than informal ones, but not always. You might report information obtained from a business trip or progress on periodic activities or on a long-term project using a set format—and yet be quite brief. Following are discussions of the specific components you can include in both short and long reports.

Components

Short Reports

Trip

If sent by your organization to visit clients or attend conferences or conventions, you might be required to report either formally or informally on your activities. In a trip report, you can discuss these activities under headers with specific topics, or you can organize them by date.

Activity

Similarly you can summarize other activities in short reports that you submit periodically. Here you might list these activities with a summary of the work completed for each and arrange them chronologically. Figure 12-1 provides an example of such a report that covers two weeks.

ACTIVITY REPORT

March 15, 200X

Purchased new equipment for office
    Researched information about five products.
    Conducted interviews with vendors about discounts, delivery, and warranties.
    Reviewed budget with manager.
    Prepared preliminary report.

Produced brochure
    Worked with artist to create sketch.
    Reviewed text from writer.
    Scheduled photo shoot.

Updated style guide
    Reviewed earlier guide.
    Scheduled meeting with dept.
    Reviewed the guides from competitors.

March 22, 200X

Purchased new equipment for office
    Met with two vendors for updated proposals.
    Visited the site to see products.
    Wrote the final report with recommendations.

Produced brochure
    Met with a photographer and reviewed photos.
    Conferred with the artist and writer.
    Arranged the printing.

Updated style guide
    Set up a style committee at the dept. meeting.
    Set a schedule for meetings and contributions.
    Began writing the introduction to the guide.

Figure 12-1: Sample activity report.

Progress

During the course of long projects, you might be asked to submit a progress report. Like an activity report, you submit it periodically; but unlike an activity report, you focus on one specific project rather than a series of them. In this report you summarize what has been done as well as what needs to be completed. Following are the basic sections that are found in most progress reports:

Introduction: Usually, progress reports begin with a short section in which you state the *purpose* of the project. Also, you give the background of your work leading up to the project. Overall, this section provides a summary of what the project is about.

Work Completed: In the first section of the report's body, you provide a summary of the tasks that you've already finished. For each task, you give the date on which you completed the task along with details about it.

Work in Progress: In the next section, you describe what task or tasks have begun but have not yet been completed and provide details about why they are still incomplete.

Work to be Completed: In this section you list the tasks that you've not yet started, but are part of your original work plan or proposal. For these tasks you need to provide details, and give the dates you expect to complete them. Often these dates will be different from those in your original plan, and thus this report updates the schedule for your project.

Problems or Unexpected Complications: If you've encountered any difficulties that might change your original schedule, you want to note them and explain briefly why they arose. Here you can explain why you need to add tasks or change dates.

Figure 12-2 on the next page shows a brief progress report. Notice how it provides an overview of the project and describes what work has been done as well as what still needs to be done.

Long Reports

Informational

The informational report summarizes a variety of activities within an organization. An example would be an annual report that includes accounts of earnings and expenditures as well as information about the current Board of Directors. Other reports describe the findings of committees on a specific topic or the results of surveys. Because informational reports are so varied, their components and structure differ. Their organization, whether by topic or chronology, depends on the type of material you have.

Analytical

Although analytical reports have different components, you usually structure them so the reader arrives at conclusions based on the information reviewed within them. In some, you might review the problem or problems in detail and then recommend solutions, while in others you would compare possible solutions or a solution based on set criteria. In these reports, you often analyze information to show if a proposed solution is feasible or to explain which alternative is most feasible. The following are some of the major components you find in such a report and the order in which they appear:

Preparatory Parts: Before the actual report begins, you need to include a letter (or memo) of transmittal, title page, table of contents, and list of illustrations. You can use small Roman numerals (for example, iii, iv, v) to *paginate these pages beginning with page iii.* Although you don't put page numbers on the letter or memo of transmittal (*page i*) and title page (*page ii*), you, nonetheless, count them in the pagination.

Letter or Memo of Transmittal: For a report for a customer or client outside the organization, you can send or "transmit" it with a cover letter. However, for an internal report for a colleague or supervisor, you would attach a memo. This letter or memo acts like a large Post-it® note in which you tell the reader what you're sending and why. Unlike a note, though, it's a formal document that highlights important parts of

---

To:	Prof. Lehr
From:	C. Daniels
Date:	October 14, 200X
Subject:	**Progress Report on the Feasibility of Opening a Discount Brokerage House in Forkland Mall**

PURPOSE

With the number of working Latinos in America on the rise, there is no outlet to aid them in planning for their future. The Latino population ranks among the lowest of cultural groups in the United States who invest in financial markets. A company that satisfies the financial need of the Latino population would be beneficial to both the community and to the brokerage organization. The objective of this study is to see whether it's feasible for a brokerage company to target Latinos in the greater Northeast Plainville region by opening an office in Forkland Mall.

WORK COMPLETED

October 10	Devised and copied a questionnaire to survey Latino shoppers at Forkland Mall.
October 13	Interviewed Maria Rodriquez, press secretary for Univision Channel 8 about the likelihood of attracting Latinos as potential clients for a brokerage house.

WORK TO BE COMPLETED

October 16	Interview Luisa Santiago, President of the Greater Plainville Latino Society (GPLS) to determine the overall attitude of Latinos to the idea of investing money in financial markets and to gain a sense of the best corporate image to use; give copies of questionnaires to be distributed to GPLS members.
October 17	Interview Leonard Smith, manager of Forkland Mall to discuss issues such as cost of rent, term, and other items.
October 21-24	Conduct surveys in Forkland Mall.
October 25	Complete and analyze data from surveys to find answers to unknown problems that arose from survey.
November 1	Conduct interviews with various charter members of GPLS to gain an understanding of how to direct the corporate image towards Latinos.
November 8	Analyze findings and begin first draft.
November 13	Submit revised draft.
November 20	Meet instructor to review the draft.
November 22-24	Write and submit a revised report.

PROBLEMS	At this point, everything is on schedule. I have encountered no problems.

Figure 12-2: Sample progress report for a feasibility study on opening a discount brokerage office to attract Latino clients in a large urban shopping mall.

the report to which you want to call attention. Figure 12-3, a Memo of Transmittal, for instance, highlights the findings in a study to determine the best ways to recruit workers for a college's food service department.

Title Page: The title page includes the name of the report, author's name, name of the person who is receiving the report, and date of submission. Type this information with a font larger than that used for body text and center it on a page. See the sample title page in Figure 12-4.

Table of Contents: The table of contents follows on the next page with the names of the topics—usually the first-level and second-level headers within the report—listed in the left-hand column with their respective page numbers listed across from each of them in the right-hand column (Figure 12-5). You can generate the text for this page in Microsoft® Word, using the "Reference" tab, and clicking onto the "Table of Contents" ribbon. You can also create it using the "Special" pull-down menu in Adobe® FrameMaker®.

To:	Prof. Lehr
From:	Leanne McMillan
Date:	October 16, 200X
Subject:	**NEW RECRUITMENT & MAINTENANCE PROGRAMS FOR LARSON UNIVERSITY FOOD SERVICES**

Here is my report on the study that was approved earlier this semester. I researched new recruitment tactics and new maintenance techniques to be applied to the university's Food Services Department.

This study took a slightly different turn from that which I had intended or expected. Initially, I believed that the members on the Food Services Board were not doing enough to get the word out about their employment opportunities. I struggled, along with two other student supervisors, to find students to fill abandoned positions, and assumed that our hardships and failures arose because we were not advertising our opportunities sufficiently. What I discovered in this area was that most of the problem developed because of personal issues with students.

My original suspicion, however, that supervisor/employee communication needed to be strengthened (and hasn't been mainly because of supervisory slacking) was confirmed. Student employees felt unsure of many of the rules and regulations they were expected to follow. My fellow supervisors and I never held a meeting to explain job duties and policies explicitly.

These findings suggest that fewer employees would quit if taught effective time management skills, and that communication could be strengthened with regular meetings and evaluations by supervisors and feedback by employees about their concerns.

I am grateful to my employees for their cooperation, patience, and honesty. I also appreciate the eager, enthusiastic encouragement and efforts of the Food Services' management. Their knowledge has been invaluable.

Thank you for the opportunity to conduct this research. It has really enriched my leadership experience, and will surely serve as great motivation in future projects.

Figure 12-3: Memo of Transmittal for a feasibility study on recruiting student workers.

Whether you create a table automatically in Microsoft® Word or FrameMaker® or create it manually, you'll want to decide whether you want **leader lines** (the spaced periods) between the topics and page numbers. You also want to ensure that all topics and numbers are aligned correctly under each other. Even if you create a table automatically with software, you sometimes need to make manual adjustments. Because the word "Table" is unnecessary, whatever way you create your table, you'll need to title it just "Contents."

List of Illustrations: This list of figures and tables appears on the page after "Contents" and consists of all of the illustrations and tables in the report. If there are several tables, you can list them separately; however, if there is just one or two, you can put them together with the figures. You can generate this list the same way you generate the content pages in Microsoft® Word or FrameMaker®. Figure 12-6 shows such a list with individual figures and page numbers.

**Expansion of the Communication Center:
A Feasibility Study**

Prof. Mary Thrugood

**Louis DeFranco
April 30, 200X**

Figure 12-4: Sample title page for an analytical report.

Contents

Figure 12-5: A Contents page for a feasibility study.

ILLUSTRATIONS

Figure 12-6: Sample List of Illustrations.

Introduction: You begin the pagination of your actual report with Arabic numbers (for example, 1, 2, and 3) in this introductory section. Here you include several sub-sections—the Problem, Purpose of the Study, Background, Scope and Limitations, and the possible Alternatives or solutions and the Criteria used to evaluate them. Figure 12-7 shows the first three subsections in an Introduction to the feasibility report on recruiting student workers.

Problem: The Problem section consists of a paragraph or paragraphs that state your dilemma—the situation that led you to conduct the analysis or study.

Purpose: Often combined with the Problem section, the Purpose states the reason why you conducted your study.

Background: The Background section consists of one or more paragraphs that allow you to elaborate on the Problem. Usually, you go back in time to summarize distant causes and describe them chronologically.

Scope and Limitations: Scope and Limitations reports on the parameters of your study and why they existed. In it, you explain what you included and what you left out along with the rationale for the decision to limit the study to particular points. Often you list the **alternatives** and **criteria** that you chose and give a brief blurb describing each, as in the outline below:

> **Alternatives**
> - Alternative 1 - *a brief statement describing it*
> - Alternative 2 - *a brief statement describing it*
> - [Alternatives 3, 4, or 5, if desired - *a brief statement describing each of them*]
>
> **Criteria**
> - Criterion 1 - *a brief statement describing it*
> - Criterion 2 - *a brief statement describing it*
> - [Criteria 3, 4, or 5, if desired - *a brief statement describing each of them*]

Figure 12-8 shows the alternatives and criteria for a study on the feasibility of establishing an online segment of a florist business.

Methods: Methods list all your sources arranged as primary or secondary sources in some logical order—usually by specific type, for example, interviews, surveys, and e-mail (see Figure 12-9).

Results: In the Results section, you provide the data for each alternative with the criteria. Here you either list the information in columns or write it in paragraphs. Figure 12-10 shows the first two pages of the Results section of a feasibility report on an online floral business with the information for each criterion given in paragraph form for three alternatives: Ftd.com, Floraplex.com, and a Do-It-Yourself site.

Conclusion: In the Conclusion section you interpret the information you provided in *Results*, and summarize your findings— for example, you say that Alternative 1 is the largest or fastest or whatever the different criteria show.

Problem

After the first few weeks of each academic semester, many new student employees within Larson's Food Services Department quit their jobs. Communication between remaining workers and management weakens. Supervisors have trouble replacing lost employees. They also have trouble rebuilding lost lines of communication.

Purpose

The purpose of this study was to determine the feasibility of initiating new recruitment techniques and maintenance programs to the Larson Food Services Department to lower the number of students lost and improve relationships between employees and their higher-ups. The idea was that if more students were hired each semester, and stronger connections were built between them and their supervisors, fewer workers would quit.

Background

The Food Services Department at Larson is an integral component in many students' daily lives. Residents living in the dormitory building are required by the University to purchase a meal plan, granting them a designated number of meals each week. Often, even though it's not required, students who have moved into campus apartments or townhouses also continue to utilize meal plans.

Food Services not only provide nourishment for students—the Department also creates multiple job openings, both budget and work-study, for undergraduates seeking on-campus employment. Numerous positions such as waiters/waitresses, hosts/hostesses, serving line attendants, market stockers, and utility workers keep Larson's young men and women on the payroll.

At the beginning of last fall semester, I was promoted from Serving Line Attendant (SLA) to student supervisor in Larson's Food Court. My first major line of duty was to interview and hire incoming freshmen from the class of 2009 to fill our available SLA and utility positions. I spoke with approximately 25 students and welcomed them all on board to join us. Of that group, only 15 remained with us for the entire semester. For a couple of weeks, 5 students put in hours before quitting, while the other 5 never contacted us again.

As for the students who remained, they were permitted to work every day of the week with the exception of Fridays. They were required to work from 4:30 p.m. to 8:00 p.m. on Monday through Thursdays, a total of 3 1/2 hours a night. Weekend shifts were the same length of time, but hours ran from 11:00 a.m. to 2:30 p.m. Most students worked 2 out of 6 days a week, a minimum of 7 hours. On any given day, there were 14 positions to fill. With only 22 employees (7 of whom were returning workers) in rotation throughout the week, the supervisor found themselves shorthanded by an average of 5 workers each night. This shortage displeased the supervisors, student managers, full-time Larson-employed Food Services workers, and customers who were often pressed for time between one class and the next.

In an attempt to fill these open positions during the first semester, word of mouth was tried. This resulted in the hiring of one worker who still remains, and another who promptly quit. For the second semester, "Need a Job?" flyers were posted in the Union to draw attention, but to no avail. The workforce did not increase in size, and recruitment procedures failed.

In addition to these failed recruitment procedures, supervisors had difficulty channeling information to and keeping contact with the student employees. The new hires attended a group introductory session during the first week of school, where their main duties were briefly described to them. After that, no further follow-up meetings were held. Periodically throughout the year, it was discovered that the students were unaware of the full extent of their responsibilities and of the exact times they were to clock in and out with their timecards. Also, even though students were given multiple updated schedules and lists of telephone numbers, they did not always show up for their assigned shifts, and some did not call supervisors to notify them of their absence. Furthermore, students did not follow the appropriate practices for obtaining a substitute, which is expected of all students who know ahead of time that they will be missing a particular day. Finally, break times were stretched out and abused by several workers.

The combination of these problems prompted a look into the recruitment and maintenance practices of the management structure within Food Services to see what improvements could be made.

Figure 12-7: Introductory sections—Problem, Purpose, and Background.

Recommendation(s): In Recommendations, you state the decision that logically arises from the interpretations in your *Conclusion*. Thus, after analyzing the statements in the *Conclusion*, you recommend the adoption of one or more alternatives. Figures 12-11 and 12-12 provide examples of these two sections.

Supplementary Materials: At the end of the report, you can include additional materials such as a listing of sources or appendices.

Bibliography: For an academic report, your instructor might require a listing of sources you consulted. These sources can appear in any format (APA, MLA, or CSE, for example) as long as the format is consistent. Check with your instructor before creating such a listing, and see Appendix C for examples of MLA and APA formats.

Appendices: Often reports include additional materials like maps, brochures, oversized charts, and letters—materials you mentioned in the report but don't necessarily put there. You can group these materials by their types (for example, maps or interviews), and label them in alphabetical order (for example, *Appendix A: Maps, Appendix B: Letters, Appendix C: Interviews, and so on*). Appendices are often paginated with capital letters designating a particular section followed by hyphens and consecutive Arabic numbers (for example, A-1, A-2, and A-3).

ALTERNATIVES

Alternative #1

Ftd.com is an e-commerce Web site service that is a subsidiary of FTD Florist's Transworld Delivery, Inc.—the largest delivery floral company in the world. Ftd.com has been on the Internet since 1994, providing advertising and promotion for the retail flower shop. Ftd.com offers an opportunity for flower shop owners to be established with the leading floral name. Its services provide adequate exposure for the flower shop Web site as well as dependable service for the online shopper.

Alternative #2

Floraplex.com is an e-commerce Web site service that provides 24-hour online sales service for wholesalers to retailers. Also Floraplex.com provides solutions for the worldwide floriculture industry; opening up global trade, offering open communication between business owners, and providing industry information and education.

Alternative #3

The **Do-It-Yourself Web Site** is the alternative Web site solutions for the flower shop owner who wants to make an initial investment without relying on services supplied through commercial Web site services.

CRITERIA

Exposure on Search Engines

When considering a Web site, Mr. Anthony needs to find the best avenue to make sure that the flower shop's Web site has ample exposure to existing customers and potential customers. After all, what good is a beautifully designed site if it's difficult to find on the Web? Having an easy-to-find Web address is the best way to establish an online identity and the quickest way to join the e-commerce revolution.

Secure Online Ordering

The second important criterion taken into consideration is securing online ordering. Many new Internet shoppers worry about credit card fraud and personal information privacy.

Figure 12-8: Alternatives and Criteria for a feasibility report on starting an online floral business.

Writing Text

Using Tense

For both information and analytical reports, you use present or past tense, depending on whether you explain what now exists or what happened earlier. If you report on a completed feasibility study, for example, you obviously use past tense. However, when you give your recommendations or solutions, you want to use present or future tense.

METHODS

PRIMARY RESEARCH

In order to evaluate the different e-commerce Internet service that are available to retail flower shops, the initial research began by surfing the Web to obtain an idea about the types of Web sites local flower shops are utilizing. Also, contact telephone number and e-mail addresses were noted for representatives from various types of e-commerce Web services that are available. Along with e-commerce Web services, contact numbers for local flower shops that constructed their own Web sites were noted.

Telephone Interviews

The following representatives were contacted:
- Herb Roth, a Ftd.com representative in the local area
- Nick Sabatino, a Floraplex.com representative for the local area
- Chris Drummond, a representative for a local flower shop with the Do-It-Yourself Web site

Also telephone interviews were conducted with three local flower shop owners for feedback on the positive and negative aspects of involvement in e-commerce business. The following local flower shops were contacted: Sally's Flower Shop, Plaza Florist, and Martier's Florist.

Personal Interviews

An informal interview was conducted with Stosh Matela, owner and operator of a small wholesale flower business. He makes daily deliveries to many retail flower shops; therefore he has knowledge about which flower shop owners are pleased or disappointed about incorporating online e-commerce business.

Surveys

About 20 people of various ages and backgrounds completed a survey so that I could obtain information about what types of customers currently shop online. These surveys encompassed questions concerning current online shopping habits and questions concerning online features customers would want to use for easier access of shopping online for flowers. The survey also included anyone who did not shop online to obtain information about the possibility of shopping online in the future.

SECONDARY RESEARCH

Various articles written and published about e-commerce Web sites such as those in *The New York Times* and *Florist Magazine* were consulted for guidance. The information collected from the newspaper and magazine sources helped to identify specific information for consideration when undertaking the task of selling merchandise and service online.

Figure 12-9: Sample Methods section for a study on starting an online floral business.

RESULTS

1. FTD. com

Visiting Ftd.com's Web site supplied sufficient information about available services and an interview with Herb Roth, a Ftd. com representative, also helped to supply sufficient information. In addition, the owner of Sally's Flower Shop provided first-hand information about her flower shop's experience with an e-commerce Web service provided by Ftd.com. Her business has been online for the past two years, and her online sales continue to increase. Another viewpoint to support Ftd.com's service was the interview with Stosh Matela, who indicated that most of the flower shops in the area that are online through Ftd.com report an increase in sales coming from online purchases.

Exposure

Because FTD is the largest floral company in the world, most floral consumers would automatically relate to FTD when ordering flowers; therefore Ftd.com would be easily located on the Internet. Ftd.com advertises on well-known search engines such as MSN as a featured store on the Flowers and Gifts link; therefore visibility for a small retail flower shop ranks high for online shoppers.

Security

To assure online shoppers of safe transactions and privacy protection, Ftd.com guarantees a 100% encryption service. Ftd.com uses Netscape's Enterprise Server, which is among the best Internet security software in the world.

Cost

The cost for Ftd.com's e-commerce Web services is $500.00 to set up the Web site, $50.00 per month for maintenance, and $2.00 for every transaction.

Maintenance

Ftd.com fees include maintenance for the Web site.

2. Floraplex.com

Visiting Floraplex.com's Web site supplied sufficient information about available services and a phone interview with Nick Sabatino, a Floraplex.com representative also helped to supply information.

Exposure

Floraplex.com is new to the e-commerce flower industry and has the ability to provide up front exposure for retail flower shops on the search engines through print and Web advertisements. Floraplex.com advertises in floral publications such as *Floral Magazine* and *Professional Floral Designer Magazine*.

Figure 12-10: Results section with information about each criterion for the three alternatives.

Security
Floraplex.com provides a secure area for commercial transactions to insure punctual payment and adheres to all applicable United States federal and state law related to the collection and use of personal information.

Cost
Floraplex.com's e-commerce Web service charges no fee to set up the site and a 5% fee per transaction.

Maintenance
Floraplex.com offers 24-hour technical support, but does not offer maintenance of the e-commerce Web sites. All maintenance is the responsibility of the flower shop owner.

3. Do-It-Yourself Web Site

Chris Drummond, the owner and operator of Plaza Florist (www.plforist.com), a local retail flower shop, offers valuable information for Mr. Anthony, if he is interested in designing, implementing, and maintaining his own Web site.

Exposure
There are various avenues to get exposure for a Do-It-Yourself Web Site, including online registration services available for a membership fee of approximately $35-$70. These services offer to provide search engine exposure by registering the businesses' Web addresses with many of the popular search engines.

Security
The Do-It-Yourself Web Site business owner commonly uses VeriSign.com, a leading provider of Internet trust services. VeriSign.com claims to have a secure server, and all information is encrypted to protect the online shopper.

Cost
A successful Do-It-Yourself Web Site will invest approximately $10,000 per year. The total cost includes e-commerce software packages, consultants, technical support, memberships in the various Web site registration services, and Web site designers.

Maintenance
There are three different ways to maintain a Do-It-Yourself Web Site. All three ways involve either time or cost:

1. the flower shop owner can learn about the software and invest the time involved in updating the site;

2. the flower shop owner can hire a Web designer for approximately $65 an hour;

3. the flower shop owner can purchase a user-friendly maintenance software program to enable the staff to make changes. The initial cost is several thousand dollars.

Figure 12-10: Results section with information about each criterion for the three alternatives (*continued*).

CONCLUSION

1. Ftd.com

Ftd.com provides an e-commerce Web site service with excellent exposure for two reasons: Ftd.com is registered with many search engines, and FTD is a common flower industry trademark known to many people around the world. A secure encryption service is guaranteed because the service utilizes one of the best Internet security software programs in the world. The cost is pricey, but the cost is equal to the benefits provided to the flower shop owner. The no-maintenance feature of a Web site available through Ftd.com is reflected in the cost.

2. Floraplex.com

Floraplex.com claims to provide adequate exposure on the Web and in print. A secure area for commercial transactions is offered to the retail flower shop owner and the online shopper. Although the cost is a fraction of the cost compared to the Ftd.com, the fact that the service is not yet well-known presents a high risk to the flower shop owner who depends solely on Floraplex.com to increase sales. All maintenance is the responsibility of the owner.

3. Do-It-Yourself Web Site

Online Web site registration services that provide search engine exposure are available for approximately $35-$70. Accredited Internet security services are available to assure safe online sales. The time and cost involved in the maintenance of a Do-It-Yourself Web Site are expensive but allow the owner to make necessary changes directly and quickly. The initial cost is high, about $10,000 per year. Even though it's a gamble, a Do-It-Yourself Web Site looks promising in the long run.

Figure 12-11: Conclusion section of a feasibility report on starting an online floral business.

RECOMMENDATIONS

Out of the three e-commerce Web site services that were examined in this report, Ftd.com is the best choice for Mr. Anthony's retail flower shop business. The familiarity of the FTD name alone would make Ftd.com a wise choice. Ftd.com would be an excellent choice for David Lautt Florist to get exposure on the Web, to guarantee his customers a secure encryption service, and to provide a maintenance free service that is highly beneficial for Mr. Anthony at this time. Also Ftd.com offers a reasonable cost for the services offered.

However, I do suggest that a Do-It-Yourself Web Site could be another alternative in the near future. With Mr. Anthony just beginning an online business, a Do-It-Yourself Web Site would be too much of a gamble. Once Mr. Anthony gains the knowledge and experience of operating an online business, I recommend that he should conduct another feasibility study to consider a Do-It-Yourself Web Site alternative.

Figure 12-12: Recommendation section of a report on an online floral business.

Writing Text (*continued*)

Maintaining Parallel Structure

In crafting a successful report you need to maintain parallel structure throughout. This parallelism particularly applies to grammatical consistency in the construction of headers and lists. All headers at the same level need to be the same grammatical structure: for example, you need to create headers that are all nouns, noun phrases, verbs, or verb phrases. Similarly, you want to construct lists where all items are complete sentences, clauses, phrases, or single words. Consistency on every level is crucial for readability.

Creating Headers

As much as possible, you want to create headers that actually say something specific about the content that follows. Avoid generic headers, and instead find words that can convey meaning precisely. If you use generic headers such as those used as examples in this book (for example, *Introduction, Conclusion, Recommendations*) be sure that the text that follows contains specific information. If possible, use gerunds for headers rather than nouns to convey a sense of action. For instance, write *Creating Graphics* rather than just *Graphics*.

Designing Pages and Using Graphics

Starting First-level Headers on Separate Pages

Major sections, those beginning with first-level headers, should each begin on a separate page, even if a large amount of extra space remains on the last page of the previous section. Like chapter headers in a book, your major sections need to be prominent.

Using Headers as Graphic Elements

Headers in reports need to be created not only so that they are parallel grammatical structures, with their names identifying the specific content that follows, but also so that they work visually as graphic elements on the page. To create visually attractive headers and increase readability, consider carefully the typefont you choose as well as its size, weight, and posture. Make sure you begin with a large enough font size for your first-level headers, so that you'll have sufficient contrast with decreasingly smaller sizes for all the sub-headers that follow.

Using Bulleted Lists

Whenever possible, break up long sentences and summaries of information in lists, and use bullets. Lists with bullets allow readers to grasp more quickly what you're trying to convey. Because bullets come in a variety of sizes and shapes, using different ones for different types of information not only makes the information more accessible, but also identifies for the reader its place in the report's overall structure.

Using Infographics

Most reports include infographics—charts and tables especially. These graphics convey information as clearly—if not more clearly—as text does. For an explanation of the different types of charts and their uses as well as how to integrate them effectively into your reports, see Chapter Six.

Checklist

	Do You Know	Yes	No
1.	how reports can be classified?		
2.	what Trip and Activity Reports are?		
3.	what information is usually included in a Progress Report?		
4.	why you want to include any problems or complications in a Progress Report?		
5.	what is the difference between informational and analytical reports?		
6.	what sections are included in the Introduction to a formal feasibility report?		
7.	what different types of research you might use and include in a report?		
8.	why the Results or Data section lists just the information you found for each alternative and criteria?		
9.	how the Conclusion relates to the Results section?		
10.	how the Recommendation or Recommendations logically derive from the Conclusion?		
11	what supplemental parts you might include in your report?		
12.	how to label Appendices?		
13.	how headers can work as graphic elements?		
14.	why you want to include lists in your report?		
15.	why you want to include lists in your report?		

Exercises

1. Give examples from your experience of formal and informal reports that you've written or received.

2. Classify the following situations as those that might result in either an informational or analytical report:
    a. an accident
    b. an inventory of a store's consumer goods
    c. a plan for solving the problem of parking on campus
    d. a visit to a client's site
    e. professional activities for a specific year
    f. a comparison of several possible sites for a new stadium
    g. a plan for a new major in a college

3. Write a short information report either updating a professor about your progress on a long-term project or summarizing activities on various projects during the last two weeks.

4. Create headers for a long paper or report you've previously written; and using Microsoft® Word, FrameMaker®, or another software program, generate a table of contents for it. What adjustments do you need to make when the table is generated?

5. While much of the information in a feasibility report appears in the Results section, some might argue that this section is not as important as the Conclusion and Recommendations. Do you agree or disagree?

6. Do such documents as maps, transcripts of interviews, copies of letters and e-mail always belong in the Appendices? When might you place such materials in the report where they are referenced?

7. What is important in choosing typefonts for headers? What should you consider when selecting type for different levels of headers and body text in a report?

8. What role do bullets and lists play in the layout of a report? What can you do to vary them?

9. Evaluate the short report on the next page (Figure 12-13) for both content and appearance. Is the information being conveyed clearly, and is it presented in a visually attractive format? What suggestions might you make to improve it overall?

10. Write a feasibility report on a study approved by your instructor that has at least two alternatives and two or three criteria. Include at least three original graphics.

To: Professor Lehr
From: K. FitzMaurice
Date: April 18, 200X
Subject: **Progress Report on the Feasibility of Constructing a Darkroom on Campus.**

PURPOSE

Larson University prides itself in being a Liberal Arts school. Unfortunately, Larson is the only area college that does not have a darkroom/photography lab on its campus. Adding a darkroom to Larson's campus would allow students of all majors to experience the arts in a way that is different from the painting, sculpting and drawing, which are currently the only fine arts courses here at Larson. Having a photography course could potentially be the deciding factor for many students looking to attend one of the five major universities in the city. It is my goal to study whether it's feasible to implement a darkroom/photography lab on campus in the renovated Holly Hall in the fall.

WORKS COMPLETED

March 19	E-mailed local universities about their photography labs.
March 21	Conducted an interview with Dr. Patricia Habler in regard to course curriculum and student interest.
March 22	Conducted preliminary research on the necessary equipment needed to construct a darkroom.
March 22	Interviewed Luke Bothman of the Student Government Association in regards to Larson construction plans for Gemini Hospital and Holly Hall.
March 30	Contacted two manufacturers of photo equipment to gather price estimates for all the equipment needed to construct a darkroom.
April 2	Devised and copied a survey developed to evaluate student interest and concern for establishing a photography lab on campus.
April 3	Distributed faculty and student surveys online and in hardcopy format.
April 17	Conducted a phone interview with Dr. Graven, University Archives, in regards to an old photography lab.

WORKS TO BE COMPLETED

April 5	Meet with facilities manager to evaluate the basement of both Olney Hall and the Unity Building as alternative sites.
April 6	Meet with Dr. Joseph Connors to review the research I have conducted and evaluate financial and curriculum issues.
April 10	Analyze findings and begin draft.
April 12	Submit a draft.
April 20	Meet for a conference with Prof. Lehr to review my draft.
April 27	Write and submit the revised draft.

PROBLEMS

Due to unseen circumstances, I have been delayed in meeting with many of these individuals. The biggest problem has been getting in touch with Larson faculty who know the future plans for construction and the reason why we do not have a photography lab already at this point in time. This was my initial question that I have yet to receive a direct answer for. At this point, I think it's most feasible to implement the darkroom when Holly Hall is being renovated. Finding alternative spaces was difficult, but I believe I have found two reasonable rooms.

Figure 12-13: A progress report for a feasibility study on constructing a photo lab.

CHAPTER THIRTEEN

Correspondence

As a professional, you'll be communicating with colleagues internally as well as with clients and associates outside your organization. Before beginning your career, you'll prepare a résumé and a cover letter to either deliver in person or send electronically to a future employer. While previously your correspondence in the workplace would have taken the form solely of letters and memos, today it occurs mainly through e-mail. Other current electronic forms—discussion groups, blogs, and wikis—function similarly in that they allow you to correspond with colleagues and customers and receive feedback from them.

Objectives

- ✓ Understand the different print and electronic forms of correspondence.
- ✓ Determine the appropriate forms of correspondence for a particular audience and occasion.
- ✓ Learn to prepare different types of résumés, and send them electronically.
- ✓ Learn rules for e-mail etiquette.
- ✓ Consider new formats—discussion groups, blogs, and wikis—as they allow you to correspond and, in some cases, collaborate with others.

Types of Correspondence

Classifying correspondence is often tricky. For example, you might classify it as correspondence used for internal and external communication, but some correspondence, e-mail for example, is used for both. You might also look at correspondence by the degree of its formality; but e-mail, memos, and letters can be both formal and informal. To classify correspondence as either print or online is somewhat arbitrary because you can print online correspondence and send letters electronically as attachments. Nonetheless, to review types of correspondence you can look at what are traditionally considered printed forms such as memos and letters. Then you can look at those forms created and transmitted electronically such as e-mail, wikis, and blogs.

Memos

Uses

Before the popularity of computers and the consequent widespread use of e-mail, memos were the workhorses of organizations and served as the main means of written communication among employees and colleagues. Today you can still use memos when you need an announcement or other information disseminated to people who might not have access to e-mail or who need a printed record of your message.

You can also use memos to accompany other printed documents and serve almost like a large Post-it® to accompany a document sent internally (see Figure 13-1 for an example of a Memo of Transmission).

Components

Memos consist of two basic parts: the address information and the message. While previously you would complete preprinted paper memo forms, now you use templates in word-processing programs such as Microsoft® Word that you can complete online.

If you need to create your own memorandum, include a line for "To," "From," and "Re" or "Subject." Then in the second part of a memo, include the message itself—usually typed flush left as a paragraph.

To: Dr. Lehr
From: Theresa Lelinski
Date: December 14, 200X
Subject: Final Report

Here is the final report for our English 303 course, Writing for Business, on the subject of whether I should choose to pursue a career writing for a major fashion magazine, or acquiring artwork for an art museum or gallery.

This study was very interesting and enlightening. My findings were much more detailed than I had originally expected, and I gathered a great deal of useful information to include. The report definitely helped me to formulate a tentative plan for what to do after my graduation.

As I had expected, the majority of my findings indicate that I should pursue a career writing for a fashion magazine. The skills necessary for such a job are more compatible with my capabilities, I would fit in more comfortably in the workplace environment, and the financial aspects are more attractive. The only downside is that I would need to relocate to New York City, and I am hesitant to move there because I would like to stay closer to home. Also, living in New York City would be far too expensive for a recently graduated college student.

Fortunately, I have come up with a solution to this dilemma. I will attempt to work for a local magazine after graduation, while saving up for the upcoming expenses of New York. These steps should help me to ease gradually into the transition.

I am especially grateful to Marcia Bowman, a friend of mine who was extremely helpful in completing this project. She provided me with an interview, helped me obtain contact information for many individuals in the field of fashion magazines, and distributed a survey to fellow students in her internship program.

Thank you for giving me the opportunity to complete this assignment. Not only has it helped me to learn about feasibility studies and business reports, but it also has helped me gain information and make decisions pertaining to my future. If you have any questions about the report, please contact me over the phone, through e-mail, or in class.

Figure 13-1: Sample Memo of Transmittal accompanying a feasibility report.

A memo lacks the salutation and complementary close usually found in a letter, but you can sign your message, if you wish, or put your initials after your name in the address part.

Letters

Uses

Letters, a type of formal business communication, serve as printed records of your transactions and can be used for a variety of purposes, to convey both good and bad news as well as routine information.

Components

Inside Address and Date: Organizations usually have a letterhead with their address. However, if your organization does not have one, you need to supply an address—the number and street on the first line; the city, state, and zip code on the second line; and the date on the third line. If a letterhead address exists, then supply the date alone.

Addressee: Next, type the address of the person to whom the letter is being sent—here with three lines, one for the name of the person and title (if long, put on an additional line), one for street address, and one for city, state, and zip code.

Salutation: Start by addressing the person, if he or she is known, by name. In formal letters, begin with titles such as Mr. or Ms. If the name is unknown, try to find the name of the person's position (for example, sales manager or customer service representative) rather than use such generic or outdated salutations such as "To Whom It May Concern" or "Dear Sir" or "Dear Madam." (Would you really address anyone in person as Madam?)

Body: In this part of the letter, you begin the actual message. Paragraphs are often short in letters and can follow a direct or indirect structure depending on their purpose, as explained later.

Complimentary Close: Professional letters usually conclude with a brief phrase such as "Sincerely," "Sincerely yours," "Cordially," or "Respectfully." With the increased emphasis on maintaining an informal tone, you might want to write a more personal closing, but generally you want to close with "Sincerely."

Signature and Name: Your typed name should appear four lines after the complimentary close; with your signature in between the two. Occasionally people sign only their first name, though the complete name is printed beneath it. However, for business letters, your signature should match the printed name below it.

Optional Components:

Attention Line: Sometimes, when you want to direct your letter to a specific person within an organization or department, you include an attention line addressed to that person in the Addressee section. The word "Attention" followed by the person's name appears after the organization's address with a space in between.

Enclosure: If you're attaching another document—a report, résumé, or contract, for example— you want to type the word "Enclosure" or "Enclosures" at the bottom of the letter to let the recipient know that such a document is also enclosed.

Copied: If you're copying someone by sending that person the same letter, you need to acknowledge the copy at the bottom of the page with either "c" or "cc" (originally used for "carbon copied") and the name of the person copied.

Types of Letters

Some letters that you write will announce good news: offering employment, granting credit, or congratulating someone on receiving an award. At other times you'll have to send bad news: denial of credit, a rejection of an application, or a refusal to grant a request. At other times you'll need to send routine correspondence such as letters of inquiry, transmittal, and employment. Descriptions of these types of correspondence follow.

Inquiry: Before e-mail became widespread, if you needed information you would write a letter requesting information. However, today you would more likely send an e-mail requesting it. Nonetheless, there are occasions when you would want to send a letter. Figure 13-2, for example, is a letter a student sent requesting information for a feasibility study from a dean at his university. Note the various parts of this letter—the request for information, the reason for the request, and the offer to provide, in return, information or something else that the recipient might want—in this case, a copy of the report.

3342 N. George St.
Pittsburgh, PA 18922
February 20, 200X

John L. Rodin, Dean
Student Affairs
Larson College
1600 W. Oaklyn Ave.
Pittsburgh, PA 18903

Dear Dean Rodin:

I am a senior conducting a feasibility study to determine the cost and practicality of making certain renovations on campus to accommodate students with physical challenges. I am writing to ask if you could help me find information related to my study.

As a physically challenged student, I have noticed a number of improvements made over the past year that have increased accessibility; and while I appreciate these improvements, I believe that other areas exist where more improvements might be made.

Particularly, I am trying to determine whether additional improvements in the Student Union would be cost effective and what other criteria should be considered in deciding to make these improvements. Therefore, I am seeking answers to the following questions:

1. What is Larson's official policy concerning handicap accessibility?
2. What amount of funds does the university budget allocate for renovations to improve accessibility?
3. What are the criteria used in deciding whether to make specific renovations, and what steps are involved in completing them?

I would be very grateful if you can answer these questions or provide other information related to them. Your assistance will help me complete my study, and I would be glad to send you a copy of my final report so that you can review both my research and recommendations.

Sincerely,

John Williams

Figure 13-2: Letter of Inquiry.

Transmittal: When you need to send a document to a client or someone else outside your organization, you may want to enclose a letter explaining its purpose and highlighting its contents. Like the memo in Figure 13-1, this letter is one of transmittal. Like the structure of the sample memo, the first part should state what you're writing about. Then in the middle part, you should highlight information from the enclosed document. In the final part, you can acknowledge outside help that made its completion possible or add a polite statement—usually an offer to supply more information or answer questions about the document's content.

Employment: Many positions require the completion of an application, but if you're answering an advertisement for a vacant position or applying for one at a particular company, whether or not a position is currently vacant, you want to send a résumé with a cover letter. Whereas a résumé lists all your qualifications and experience, the accompanying cover letter usually calls attention to specific experience and skills that you have that match those the employer is seeking.

Cover Letter

INTRODUCTION: In writing your cover letter, begin by stating what position you're applying for and—if you're responding to an ad—where and when you learned that such a position is available—for example, in a newspaper on a specific date, on a web site, or from a friend. You might end this first paragraph with a statement indicating that you have the skills and experience needed for the position and then proceed in the next paragraph to tell the employer about them. See Figure 13-3 for such an introduction.

BODY: In this next paragraph you usually provide specific examples of experience listed on your résumé that serve as evidence that you possess the requested skills and experience. In preparing to write this paragraph, you might underline key words that appear in the advertisement to ensure that you use them. For instance, if the advertisement asks for someone with good communication skills, write that you have good or perhaps superior communication skills and give examples of activities that have shown them. You also would want to tell about your writing ability and what types of writing you've done in school or professionally. Of course, you want to avoid either exaggerating your achievements or downplaying them.

CONCLUSION: In the final paragraph, you want to reiterate your interest in the position and ask for an interview. You do so by offering to provide more information or references if needed and by offering to meet with the employer at his or her convenience. So that the employer can reach you immediately and set up such a meeting, conclude by giving your phone number or e-mail address and mentioning a convenient time when someone can reach you. Even though your contact information is available on your résumé, you want to repeat it here in the closing of the letter so that it's readily available.

Résumé

A résumé offers an employer an overview of your abilities and background and offers you the opportunity to showcase yourself. Traditionally, the résumé focuses on your work experience—organized chronologically from the present to the past— and on your education, also organized chronologically with your most recent school listed

1400 W. Olney Ave.
Malvern, PA 19355
March 17, 200X

Mr. Adrian McHenry
BIZ Company
1400 Market St.
Philadelphia, PA 19035

Dear Mr. McHenry:

I am responding to your listing that was posted March 1 on Monster.com for a dynamic technical writer with superior writing skills and extensive experience working with software. As you can see from the enclosed résumé, I have worked with software programs for the last ten years as a technical writer and have developed excellent writing skills. Overall, I have the qualifications you are seeking.

In the past five years as a technical writer focusing on online help, I have worked on both group and stand-alone projects. As a new member of the six-person Global Documentation Team, I quickly taught myself to use the authoring tool RoboHelp 5.0. In time we purchased RoboHelp X5 enabling me to learn the new applications and share my knowledge with the other members of our team. Using this tool, I created help systems new to our company – both JavaHelp and non-compiled HTML help.

One of my goals as a technical writer was to become part of the development process. To achieve that goal, I established regular meetings with the development leaders to review the projects on which they were working. After we met regularly, I was more easily accepted at software design reviews, and my input was recognized as being valuable.

As part of a documentation team, each of us was responsible for certain projects, such as planning software or for certain areas of our large group project. I frequently assisted other teammates who were behind in their assigned software issues by completing the issues for them while they were absent or busy on other projects.

Overall I like learning new tools and methodology to broaden my knowledge and experience as a writer. I quickly learn new applications and processes and enjoy working with others. I hope to be a valuable asset to any team by assuming responsibility for projects and providing feedback to other team members.

May I arrange a mutually convenient time to discuss this position with you in person? Please call (215.405.0900) or e-mail me (wagner@nsm.com) at your convenience. I look forward to hearing from you.

Sincerely,

John R. Wagner

Enclosure

Figure 13-3: Sample Letter of Employment.

first. However, sometimes you might want to organize your résumé differently, especially if you have a gap in your work experience or your education. In such a case, you need to change the format to a functional résumé so that your skills and abilities become more prominent.

TRADITIONAL: A traditional résumé usually begins with basic contact information— your name, address, phone number, and e-mail address. Then an *Objective* and an *Education* or a *Work Experience* section follow (see Figure 13-4).

Dana Michelle Madonna
5340 Ivy Stream Road • Hatboro, Pennsylvania 19040
Phone 222-555-5555 • E-mail techwriter@someemail.com

Objective

To obtain a position where I can use my technical writing and editing abilities to create documentation that will achieve superior customer training.

Education

La Salle University, Philadelphia, PA
Graduated: May 200X, magna cum laude, B.A. in English
GPA: 3.6

Work Experience

Insurance Data Processing, Wyncote, PA: June 200X to present
Documentation Specialist
Create software user-facing help in compiled HTML, Web, and printed formats.
Edit and format training documents for uniformity and consistency.
Update document repository Web page used as an associate help resource.
Planned, scoped, and currently wrote IDP's Internal Style Guide.

Related Activities

Chair, Publicity Committee, STC - Metro Chapter: October 200X to present.
Plan and implement promotions for chapter's events and programs.
Design flyers and write emails to be distributed to organizations, other chapters, and individuals according to promotional plan.

Writing Fellow, Larson University: 200X-200X
Worked with assigned classes of students to revise and edit essays written by those students for their classes.

References and writing samples provided upon request.

Figure 13-4: Traditional Résumé.

If you begin with *Work Experience*, you should list each position you held, along with the name of the organization and the dates (usually just the years) in which you held each position. Under each of these listings, you also provide brief descriptions of your responsibilities. In writing these descriptions, begin with a strong action verb, and use specific details to explain exactly what you did.

Whether you put *Education* before or after *Work Experience*, list the school you attended most recently first, along with the dates you attended as well as any degree you received. Next, list the school you attended previously, and then any additional schools in reverse chronological order. If you've earned a GPA of 3.0 or higher, include this average along with your school. List also your major and minor subjects and any honors you received.

Other sections to include on your résumé are those that list your skills, interests, and activities as well as any leadership positions you might have held in these activities. List any offices you held—especially any elected ones.

FUNCTIONAL: If you want to emphasize your skills rather than a consistent work record, you might transform your traditional résumé into a functional one (see Figure 13-5). To make this transformation, after your contact information at the top of the page, begin with a section titled *Skills*. In this section, list the major skills that you believe you have, and support each listing with evidence from your volunteer work, activities, or part- or full-time work experience. This section takes time to develop and requires some creativity because you're analyzing what specific skills underlie your different experiences.

Follow this section with *Work Experience* or *Education,* and list your prior employment and degrees as you would list them in a traditional résumé. These sections need as much development as *Skills* does, because the whole purpose of this format is to get a prospective employer to look at your skills rather than your work experience.

FREQUENTLY ASKED QUESTIONS ABOUT RÉSUMÉS

- Should résumés include Objectives?

  Whether you construct a traditional or functional résumé, you must decide whether or not to include an objective statement. This statement is a brief description of the position you're seeking. While some people believe that every résumé should have an objective at the top, others think such a statement is unnecessary and will pigeon-hole a prospective employee who might be rejected because he or she does not fit a currently available opening.

  In a survey of 250 employers who recruit undergraduates at local colleges and universities, almost two-thirds of those who responded said that they prefer to see an objective statement on a résumé mainly because it allows the recruiter to see what position the person actually wants. However, while only one-third thought that the objective statement was unnecessary, those same respondents gave very extensive reasons for their responses. Overall they felt that such statements were too general to be valuable and that a specific objective is better stated in the cover letter. One respondent wrote that the only time the objective stands out is if it has a misspelling or some other obvious mistake in it.[1]

---

1   Survey of 250 employers who recruit at La Salle University, Philadelphia, Pennsylvania, conducted October, 2004.

JOHN R. WAGNER
1400 W. Olney Ave.   Home: (215) 555-1234
Springfield, PA 19141   Email: wagmerl@earthlink.com

SKILLS

Training   Trained clients on using company software applications.

Consulting   Consulted at client sites on application of sales and pricing in Initech
ERP software enabling more effective use of its product.

Computing   Used RoboHelp X3-X5, ForeHelp, Microsoft® Office, Visio Professional,
CVS, PaintShopPro, HTML, CSS, Macintosh O/S, and Microsoft Windows O/S

Writing   Wrote online help for a large, collaborative Windows help project
totaling over 6,500 topics.
Created, authored, and maintained help for smaller WebHelp projects. Designed and
implemented tests for help projects.
Collaborated with development managers, business analysts, and developers on
product design, software prototypes, layout, and user interface standardization.
Collaborated with developers on what information is added to online Help.
Designed cascading style sheet and customized graphics for RoboHelp WebHelp
projects to ensure a standard look and feel.
Created and maintained Microsoft Word templates that I then applied to installation
documentation to standardize them.
Clarified industry-specific terms in collaboration with translators to ensure the
completion of the process.
Produced HTML documents in text editors as well as HTML editors.
Composed tests for help projects, and then organized teammates to assist in
projects.
Developed materials on software applications to train clients and new personnel,
both in-house and on-site, thus lowering learning curve for new users.

EMPLOYMENT

199X-200X—IniteX Corporation, Philadelphia, PA

EDUCATION

199X—BA, magna cum laude, history and English, Larson College, Wilmington, DE

PROFESSIONAL DEVELOPMENT

Attended WinWriters/WritersUSA Conference in Seattle, 200X.
Board member of the Philadelphia Metro Chapter of the Society for Technical
Communication (STC).
Presented "Merging Customer-Written Help into an Online Help Project"
at Philadelphia Metro Chapter of the Society of Technical Communicators, 200X.
Attended WinWriters/WritersUSA Conference in Boston, 200X.

Figure 13-5: Functional Résumé.

- How should cover letters and résumés be sent electronically?

  Whether your résumé is a traditional or functional one, you'll most likely be expected to submit it, along with a cover letter for some positions, electronically. To do so, you can submit both as attachments to an e-mail, or else paste them directly into the e-mail. Since different browsers encode formatting differently, you need to remove the formatting first (see Figure 13-6).

- How do employers prefer to receive cover letters and résumés?

  The latter approach is actually preferred because some employers are hesitant to open attachments from senders who are unfamiliar to them, for fear of viruses

---

DANA MICHELLE MADONNA
5340 Ivystream Road • Hatboro, Pennsylvania 19040
Phone 222-555-5555 • E-mail techwriter@someemail.com

KEY WORDS
Writing, documentation specialist, English, marketing, HTML, PhotoShop, FrameMaker®, Spanish, French, oral communication, leadership, technical, training, editing, Dreamweaver®, travel, interpersonal skills, team-oriented

OBJECTIVE
To obtain a position where I can use my technical writing and editing abilities to create documentation that will achieve superior customer training.

EDUCATION
Larson University, Philadelphia, PA
Graduated: May 200X, magna cum laude
B.A. in English, G.P.A.: 3.6

WORK EXPERIENCE
Insurance Data Processing,
Wyncote, PA
June 200X to Present
Documentation Specialist
*Create software user-facing help in compiled HTML, Web, and print formats.
*Edit and format training documents for uniformity and consistency.

RELATED ACTIVITIES
Chair, Publicity Committee, STC - Metro Chapter
October 200X to present
*Plan and implement promotions for the chapter's events and programs.

Writing Fellow, Larson University: 200X-200X
*Worked with assigned classes of students to revise and edit student essays.

References and writing samples provided upon request.

Figure 13-6: Electronic Résumé.

infecting their computers. Therefore, to avoid having your résumé ignored, you want to prepare it for electronic transmission.

- How do you prepare a résumé for electronic transmission?

Take your traditional résumé, and do the following:
- ✓ Add a key word summary (20-30 words) related to your background.
- ✓ Move all text to the left as far as possible.
- ✓ Remove all bolding, underlining, and italicizing.
- ✓ Differentiate headers by using all capitals for them.
- ✓ Replace bullets with asterisks (*) or lower-case o's.
- ✓ Save the text as "rich text format."

After completing these steps, copy and paste your letter and résumé directly into your e-mail.

To ensure that you've properly prepared the letter and résumé for electronic transmission, review it by first sending it to yourself or another e-mail address to which you have access so that you can see how the document or documents actually appear. Figure 13-6 is an example of a traditional résumé that has been converted into an electronic one by following the directives above.

E-Mail

Perhaps the most popular form of correspondence is e-mail. Replacing the memo as the most frequently used type of written communication in business and industry, the e-mail message allows you to reach people in all kinds of positions quickly in a way that was previously unheard of. The democratization that e-mail provides is one of its main advantages, but the continued barrage of messages that requires constant review can counteract this advantage.

Components

E-mail is constructed very much like a memo in that it has a header section with "TO," "FROM," and "SUBJECT." Also, it has a section for your message. However, unlike a memo, it often has a salutation (*Dear* _____) and a complimentary close (*Sincerely,*); and like a letter you can include a signature. However, it has its distinctive options for sending your messages. For instance, many e-mail programs allow you to both check "return" so that you know when the message has been opened, and to designate its priority.

E-Mail Etiquette

Perhaps because of its ease of use and immediacy, you might be tempted to send e-mail without regard to some basic conventions. Some people do, in fact, ignore "Netiquette," the etiquette conventions for corresponding via the Internet. While these conventions vary, most are based on the courtesy that you extend to your readers. Following are some e-mail rules to keep in mind.

1. **Never write anything in e-mail that you would not put on a postcard.** Remember e-mail can be forwarded without your permission and can be retrieved from your computer, even after you delete it from your screen.

2. **Save e-mail for an appropriate situation.** Use it for routine messages but not for something that should be communicated in person (for example, reducing an employee's hours) or written in a letter.

3. **Adjust your style to your audience.** While you can write informally to friends, you want to write more formally to professors, employers, and other people. With friends, you can write in all lower case letters with no punctuation; with others you want to write in complete sentences—not fragments. Use conventional marks of punctuation (start sentences with capital letters, capitalize the pronoun "I," and end sentences with periods). Also, while you can use slang and incorporate emoticons in writing to friends, use plain English for everyone else.

4. **Avoid flaming**— that is, using e-mail to ventilate anger and other emotions. While you might not care at the time if someone other than your recipient reads what you've written, as Rule 1 notes, someone else could read it. Also, as with letters, you don't want to put in writing something you might regret later.

5. **Keep your messages relatively short.** Remember people don't want to scroll endlessly. If you need to send a very lengthy message, consider writing it in a word processing program and sending it as an attachment.

6. **Copy people, but be wary of blind copying and forwarding.** You want to copy people whom you mention in your e-mail if they need to be aware of what you've written. However, be wary of blind-copying others. While e-mail is not private, people should know that you've written someone else about what they are involved in. Be wary of forwarding e-mails that have been forwarded to you, especially those with a long list of recipients. If you must forward messages, erase the long list of recipients, so that your recipient can get to the actual message without much scrolling.

7. **Check for accuracy, missing words, and spelling.** When you finish writing, take a few minutes to proofread and check for the accuracy of what you're saying and for misspelled and missing words. E-mail, like other forms of communication, represents you and, if needed, can serve as a legal document, so be careful when sending it.

Discussion Groups, Blogs, and Wikis

With advances in technology, more and more people communicate with each other through the Internet —not only through e-mail but through other electronic means such as discussion groups, blogs, and wikis. While these formats might not seem like the traditional types of correspondence discussed in this chapter, they do allow you to convey information and ideas to others and, in return, receive responses from many readers.

Discussion Groups

Using Lotus Notes or other software, you can easily set up a discussion group that allows you as a technical professional to interact with colleagues or customers. Through threaded discussions, you can ask questions about a topic, propose a plan, provide a solution to a problem, or just comment on a remark from a previous contributor. While these discussions are helpful during a project, they also help you to keep abreast of innovations in your field and discuss your work with others.

Blogs

Blogs or "Web logs" are online journals that also offer opportunities for correspondence, since readers can send their comments back to you. Because they often include links to other Web sites, they become part of a larger communication network. Once set up, they are easy to update and become a means for you to post entries on a variety of subjects. For example, you can provide technical tutorials such as a brief explanation of how to fix a problem with your Mac computer or how to connect iTunes to the iTunes Music Store. Or you might blog on a more complex subject such as a content management system.

Wikis

A wiki, from a Hawaiian term "wiki wiki" meaning very fast, consists of loosely constructed web pages strung together. According to Emma Tonkin, they offer a "space containing pages that can be freely written or edited by anybody."[2] Tonkin explains that they "are not usually written directly in HTML" but rather use "a simplified system of markup." The concept of the wiki was invented by software developer Ward Cunningham in 1995, who was seeking a tool that would allow him "to trace a document's history and support revision control in order to check for previous changes."[3] The most well known wiki is, of course, Wikipedia (www.wikipedia.org) that was started in 2001 and has millions of entries that people can add to and edit. Like Wikipedia, other wikis provide records of all changes made, and thus a record of collaborations.

Ann Gentle, a senior technical writer at Advanced Solutions International, explains that wikis contain different types of information. The SplunkBase wiki (www.splunk.com/base), for example, contains large amounts of reference information specifically for IT troubleshooting. Other wikis contain knowledge-based information, where both product users and a customer support team collaborate on a list of frequently asked questions and answers. Gentle explains that wikis also have task information, such as that provided by eBay (www.ebaywiki.com) whereby users can perform different tasks as both buyers and sellers.[4] Whatever type of information a wiki contains, it offers an opportunity for many participants to collaborate and thus correspond with each other.

Writing Text

Whether writing a letter or e-mail message, posting a blog, or contributing to a wiki, you want to keep in mind your audience, the degree of formality demanded by the writing situation, and the subject matter of your text. Consider what your audience wants to read and how you might organize your message so that they understand what you're saying.

The "You" Attitude

In corresponding with anyone, maintain the "you" attitude. You don't need to continually use the word "you," but you need to keep in mind who the reader is, what the reader wants to know, and how the reader wants to receive your message.

---

2   Tonkin, E. "Making the Case for a Wiki."*Ariadne* 42(January 2005). 3 June 2007 <http://www.ariadne.ac.uk/issue42/tonkin/>.

3   Schwall, J. "The Wiki Phenomenon."(2003):4. 3 June 2007 <http://www.schwall.de/dl/20030828_the_wiki_way.pdf>.

4   Gentle, A. "The 'Quick Web' for Technical Documentation." *Intercom* September/October 2007: 16-19. < http://www.stc.org/intercom/PDFs/2007/20070910_16-19.pdf>.

Don't write: *Your payment is overdue, and we can't process your application until we get it from you.*

Write: *Please submit your payment so that we can process your application quickly.*

Notice that in the first example the words "you" and "your" appear, but the tone is accusatory with the emphasis placed on the inconvenience that the overdue payment is causing the writer— not the recipient. In the second example, the writer uses both "your" and "we," but the tone is neutral, with the emphasis placed on helping the recipient.

Degrees of Formality and Diction

Along with the "you" attitude, you want to keep in mind how formal or informal your correspondence should be. As explained earlier, letters usually require a more formal tone and diction while e-mails tend to be more informal, though they too can differ depending on the audience, subject, and situation. Whatever you do, you want to avoid traditional business jargon that many people think is necessary for a formal tone. Some examples include "per" instead of "as" or "according to" and "it is brought to our attention" instead of "we noticed." In general, write with clear, simple diction whether you're using a formal or informal style.

Structure

The organization of your text is tied to the type of correspondence you're sending. If you're sending a routine message or one that conveys good news, whether through a letter, memo, or e-mail, you want a **direct approach** wherein you state the purpose upfront in the first paragraph and elaborate on the subject in the sentences or paragraphs that follow.

However, if you're sending a message that will probably be unwelcome, you might use a more **indirect approach** and give the bad news later in the text. If you take this approach, be careful not to bury or blur the bad news. Whether welcomed or unwelcomed, you need to state the message clearly, so that there is no possibility that the recipient will misconstrue its meaning.

Designing Pages

As discussed earlier, many templates exist online or on your computer for creating letters and memos. You can personalize them as well as electronic formats for e-mail. However, with these and other types of correspondence you still need to follow some basic principles of design (for example, repetition, alignment, proximity, and contrast) and provide sufficient white space.

Design of Correspondence

With all types of correspondence, you need to be sure to leave sufficient space between the basic components. While e-mail and memos often have set designs, letters allow you to some extent to vary the design based on whether you want a block or semi-block style. Figure 13-7 shows the layout for a letter formatted in block style.

Design of Résumés

In designing a résumé, you usually place your headers in the left column and the body text across from them. While formatting your résumé, consider the following:

- Use the same typefont and size (9-12 point) for headers.
- Use a sans serif font (for example, Arial or Calibri) for headers that can be contrasted with a serif font (for example, Times Roman, Garamond, or Cambria) for body text or vice versa.
- Supply bullets for listings under headers.
- Make similar items parallel in structure.
- Keep similar information together.

When converting your traditional résumé to an electronic one, keep in mind the principles of alignment and proximity when moving your words flush left as well as the principle of contrast when using all capitals for headers and upper- and lower-case letters for body text. Figure 13-6 provides an example of an electronic résumé where these principles appear.

(Inside Address)

_____

_____

_____ (Date)

(Addressee)

_____

_____

_____

_____ (Salutation)

_____

_____

_____

_____

_____

_____

_____

_____

_____

_____

_____ (Complimentary Close)

(Signature)

_____ (Printed Name)

_____ (Enclosure or Copied)

Figure 13-7: Format for a letter in block style.

Checklist

	Do You Know	Yes	No
1.	the uses of a memo and letter?		
2.	the components of a letter?		
3.	how the content of an employment letter differs from that of a résumé?		
4.	what to include in a letter of transmittal?		
5.	what to include in an employment letter that accompanies a résumé?		
6.	the difference between a traditional and functional résumé?		
7.	how to convert your traditional résumé to an electronic one?		
8.	the pros and cons of including an objective on your résumé?		
9.	how e-mail differs from a memo?		
10.	the rules for e-mail etiquette?		
11	the advantages of using a blog or wiki as a means of corresponding and collaborating with colleagues and peers?		
12.	why the "you" attitude is important?		
13.	how the degree of formality in correspondence relates to diction?		
14.	when to use the direct and indirect approach in structuring a letter?		
15.	what to keep in mind when designing résumés?		

Exercises

1. Identify what form or forms of correspondence you'll use for each of the following.
   a. message offering a prospective employee a position
   b. plan for structuring the first part of a project
   c. a message to all employees giving information about the company picnic
   d. your ideas about how a sales campaign should be conducted
   e. a message to a colleague asking if he or she would like to meet for lunch

2. Write a letter to someone who has requested an increase in his or her credit limit either announcing that the increase will be given or denying the request.

3. Write a letter or e-mail requesting information or materials for a project you're working on. What might you offer the person as an incentive to get him or her to send you what you need? Be sure in laying out your letter that you leave sufficient white space.

4. Explain why someone might want to convert a traditional résumé to a functional one. Would someone want to create more than one functional résumé? Why or why not?

5. What can happen if you don't remove formatting from your résumé when pasting it into your e-mail?

6. Give some advantages and disadvantages of including an objective on your résumé.

7. Do you agree that you should proofread e-mail carefully and correct spelling and punctuation? Why or why not?

8. Evaluate the cover letter (Figure 13-8) and résumé (Figure 13-9) that follow on the next pages for both the writing and layout or design.

9. Find an advertisement for a position for which you might qualify either in the newspaper or online. Identify the specific skills and experience required, and give one or two examples from your background that match them.

10. Write a letter applying for the advertised position you analyzed in the previous example, and create a traditional résumé to accompany it.

11. Find a blog on a topic that interests you. What distinguishes this blog from other types of writing? What comments might you send to the blogger?

12. Go to the e-Bay wiki (wwww.ebaywiki.com), and read about how it operates as a wiki. Explain how this wiki differs from a discussion board or blog.

4000 Filbert St, Apt 7
Beaver, PA 18001

September 27, 200X

Mr. John Kenny
Real Magazine Inc.
4770 Philmont Place
Suite #5
San Francisco, CA 89006

Dear Mr. Kenny:

Bob Miller, graphic designer, has informed me that your company is seeking an experienced writer. As a Beaver University graduate with a B.A. in Communications and English, I am very confident that I would make a positive contribution to your company.

Enclosed is my résumé, and as you will see, I have had much writing experience. This summer, I have written press releases for Beaver University's PR Department, University Communications; an experiential piece for Beaver's alumni magazine, *Beaver Magazine;* and various cultural arts articles for Beaver's community newspaper, *The Spirit.* In addition to my journalism skills, I have also taken courses in technical and professional writing and have produced reports and proposals.

I am a hardworking, conscientious, and dedicated writer and would love to work for your company. You may contact me anytime on my cell at 268-985-2180.

I look forward to meeting with you shortly.

Sincerely,

Susanna Montana

Enc: Resume
Enc: Writing samples

Figure 13-8: Employment Letter.

Susanna Montana
SusannaMontana@Beaver.edu

4000 Beaver St, Apt 7 · Beaver, PA 18001

OBJECTIVE:
To obtain a paid position as a freelance cultural arts journalist

WRITING EXPERIENCE
Public Relations Director: *Real Magazine*
Cultural Arts writer: *The Spirit*
Freelance writer: *Beaver Magazine*
Freelance press release writer: *Beaver University Communications*
Staff writer: *The Collegian,* campus newspaper
Freelance Op-Ed/ Teen writer: *The Beaver Daily Record*
Fiction/Essay editor: *Fiction writing I, Advanced Fiction II, Mass Media Writing Seminar,* Beaver University
Scriptwriter: *Backstage Pass,* a campus entertainment television show
Featured fiction/creative non-fiction writer/poet: *The Grimoire,* campus literary magazine
Public relations writer: Press release and newsletter creator for *Alternative Fuel Methods Inc*
Songwriter/musician: *Late Night at Beaver* paid performer

INTERNSHIPS
Intern: NBC Olympics - Torino, Italy
Transcribed interviews, press conferences, promos, and packages for producers and reporters; recorded feeds; assisted producers in choosing sound bites for packages; fed footage to news department; researched and gathered pieces of footage for producers; shadowed the edit room; shadowed interviews; networked with Olympians
Public Relations Intern: Alternative Fuel Methods – Beaver, PA
Created newsletters, flyers, and letters for members and sponsors; telephoned Congressmen and women, and executive directors about business roundtables
Research Intern: Beaver Environmental Network – Beaver, PA
Researched environmental issues regarding local communities, national organizations, and the Federal government

LEADERSHIP ACTIVITIES
Vice president: National Society of Collegiate Scholars
Executive producer and host: *Backstage Pass,* campus television show
Co-founder & Co-president: *LaCycle,* two-year recycling project
Public relations officer: Association of Women in Communications
Member: Lambda Pi Eta Communication Honor Society
Documentary Filmmaker: Reporter and editor of *The Day in the Life of an RA/CA/RD*
Honored speaker: National Society of Collegiate Scholars
Commencement speaker: Beaver High School

SCHOLARSHIPS
Beaver University Founder's Scholarship
Citizens Scholarship
American Legion Scholarship
Veterans of Foreign Wars Scholarship
Catherine Doran English Scholarship
Greg Argenziano Writer's Scholarship
James A. Finnegan Fellowship

EDUCATION
Beaver University: 3.92 GPA, Communication/English double major, B.A. May 200X
Awards: Most Creative on Campus Award, DAR Award, Student of the Year Award, Most Likely to Succeed Award

References available upon request

Figure 13-9: Functional Résumé.

CHAPTER FOURTEEN

Guides and Promotional Materials

While working in a technical environment, you might be asked to help prepare guides for internal and external use. This chapter provides an introduction to two common guides: the style guide for employees and the quick reference guide for both employees and customers. The chapter also looks at three types of promotional materials that you might also help prepare—the flyer, brochure, and catalog.

Objectives

- ✓ Learn about style guides—their purpose, content, and benefits.
- ✓ Learn about quick reference guides—what they are and what to include in them.
- ✓ Understand the purpose and content of flyers, brochures, and catalogs.
- ✓ Understand how to integrate text and graphics within these guides and promotional materials.

Guides

Guides can range in size from one page to thousands of pages. However, when you produce them internally, you might create relatively brief documents that you can publish either in print or online. Two that are quite different but equally needed are those created to ensure consistency of style for your publications and those that provide a quick reference for a more elaborate procedure related to a product.

Style Guide

Purpose

A style guide, according to Jackie Damrau, is "a form of reference document that a technical document department uses to ensure that all its documentation has the same professional look."[1] Damrau identifies its purpose as two-fold:

1. It shows the basic concepts for producing a written or online material that has a uniform look, and

2. It embodies corporate culture and values or organizational identities.[2]

---

1   Damrau, Jackie. "Developing a Corporate Style Guide" Proceedings, STC's 53rd Annual Conference. Arlington, VA: Society of Technical Communication, 2006. 184.

2   Damrau. 184.

Such a guide serves as an agreement among you and your colleagues as to how your documentation should be appear.

Content

A style guide obviously varies in content from one organization to another; however, many style guides share similar parts. They usually include a section on writing—detailing what words to use (often with their definitions), how to use them, and how to punctuate or capitalize them in sentences. They also include information about acceptable abbreviations. Figure 14-1 shows an example of the contents page of a style guide that focuses on many of these aspects of writing.

In addition to writing, many style guides include information for formatting graphics—detailing sizes and placement of illustrations or use of a company logo, and identifying acceptable typefonts for different publications.

What to Keep in Mind

- Gain a consensus from stakeholders about what should be included within the guide.
- Appoint one person to be in charge of producing and updating it.
- Include as much information as possible about areas that might cause ambiguity and confusion.
- Keep the guide simple and brief enough so that it's not overwhelming to users.
- Determine when and how often your organization will review it for updates.

IDP Style Guide
for Technical Documentation

Contents

Abbreviations 1
Appendix 1
Brackets 2
Bullets 2
Capitalization 3
Commands and buttons 3
Contractions 3
English version 4
Notes 4
Numbering conventions 4
Required information 5
Run 5
Sentence structure 5
Voice: active, passive 6

Figure 14-1: Contents page of a corporate style guide. *Reprinted courtesy of IDP, Inc.*

Quick Reference Guide

Purpose

Another publication that you might help prepare is a quick reference guide. Often this guide appears as a laminated card, large folded sheet, or small spiral-bound booklet that guides the user quickly through a procedure involving the operation of a system or piece of equipment that would take a lot longer to read in a full-length manual.

Content

While the content of a quick reference guide varies from one operation to another, most consist of instructions or descriptions explained through text and graphics in an abbreviated form. Often they contain notes, warnings, and cautions placed prominently in this condensed, simplified version.

Figure 14-2 shows the front page of a guide for a printer that serves as a quick reference to the status of lights for the printer's operation.

Figure 14-2: Handy reference guide for operating a printer. *Reprinted courtesy of Oki Data Americas, Inc.*

What to Keep in Mind

As you would expect, the writing for such a guide must be clear and direct. It must include text that is parallel in structure and simple to follow, all within an effective visual design that includes tables or icons. For a sense of what is important in creating these guides, see Figure 14-3, an evaluation sheet produced by the Society for Technical Communication for judging publications in its annual competition. Note that it asks judges to look at whether the information included is appropriate for the guide and whether the guide uses icons, symbols, captions, and callouts effectively.

Promotional Materials

While marketing communications departments and external agencies often produce promotional materials, you might be asked to help prepare them. While your text for these materials needs to be clear and crisp, your design should attract readers as potential clients or customers. Some of the criteria used for judging promotional materials in the STC's annual publications competition appear in Figure 14-4. Note that the criteria include evaluating the effectiveness of the marketing messages, the call to action, and the overall design. Following are some of the most popular types of promotional materials.

Flyers

Purpose

Flyers are usually one-page advertisements designed for specific events and include one or more graphics, with one graphic in each being dominant. They are designed, generally, to grab the viewer's attention rather than provide an infinite amount of specific details. Because they contain dated information, they are readily disposable after the event, and relatively inexpensive to produce. Although you can print them with four colors on expensive glossy paper, usually you reserve such printing for brochures.

Content

Whether you create flyers for a trade show or a company picnic, you need to include information about the time, place, and location. While you can put this information in a subordinate place, you need to give prominence to the event itself or someone connected with it. Figure 14-5 shows a flyer advertising a local chapter's STC Conference. Notice in this example that the emphasis is on the theme of the event "Document Engineering," the keynote speaker, and his background. Although the time, date, and location are there, this information is much less prominent.

CONTENT AND ORGANIZATION					
Criteria	SD	D	A	SA	NA
Writing tone and style suit the purpose and audience					
Vocabulary and reading level are appropriate for the audience					
Organization and conventions are either inherently understandable or are explained					
Information included is appropriate for quick reference					
Strategy for organizing the information suits the purpose					
Writing is crisp and clean, with logical development of the subject matter at the right level of detail					
Technical complexity is handled effectively					
Graphic elements are positioned near the text they support					
Notes, cautions, and warnings are clearly identified, positioned appropriately, and follow conventions for their meaning					
Writing is free of gender or ethnic bias					
VISUAL DESIGN					
Criteria	SD	D	A	SA	NA
Overall design is unified and appropriate for the purpose					
Layout of page elements contributes to readability and usability					
Typography is used as an effective design element					
Typography is easy to read					
Headings are visually effective in helping readers find information					
Other navigation devices are used, as appropriate					
Graphics maintain the internal consistency of the publication					
Icons and symbols (if used) are explained and used effectively					
Graphics are suitable for the audience in tone, style, and content					
Graphics support the content effectively					
Graphics are consistently well designed, legible, and executed neatly					
Tables, charts, and diagrams are treated as graphic elements					
Captions and callouts are effective for illustrations, tables, photos, and other graphics					
Color (if used) adds to the appeal and usability of the publication and unifies its design effectively					
Size and binding are appropriate for purpose and audience					
Production materials are of appropriate durability and quality					
Print quality supports the readability and usability of the publication					

Figure 14-3: Section of judging form for evaluating quick reference guides. *Reprinted courtesy of the Society for Technical Communication.*

Content and Organization					
Criteria	SD	D	A	SA	NA
Marketing messages are clear and concise and writing focuses on conveying the messages					
Customer benefits are clear and concise and balanced with product, service, or organization features and other information					
Customer "call to action" is clear and provides all information necessary for the reader to take the desired action					
Writing tone and style suit the purpose and audience					
Vocabulary and reading level are appropriate for the audience					
Writing is crisp and clean, with logical development of the subject matter					
Technical complexity is handled effectively					
Graphic elements are positioned near the text they support					
Writing is free of gender or ethnic bias					

Visual Design					
Criteria	SD	D	A	SA	NA
Design attracts immediate attention and invites reading					
Overall design is unified and appropriate for the purpose					
Layout of page elements contributes to readability and usability					
Typography is used as an effective design element					
Typography is easy to read					
Headings are visually effective in helping readers find information					
Other navigation devices are used, as appropriate					
Graphs maintain the internal consistency of the publication					
Graphics are suitable for the audience in tone, style, and content					
Graphics support the content effectively					
Graphics are consistently well designed, legible, and executed neatly					
Tables, charts, and diagrams are treated as graphic elements					
Captions and callouts are effective for illustrations, tables, photos, and other graphics					
Color (if used) adds to the appeal and usability of the publication and unifies its design effectively					
Size and binding are appropriate for purpose and audience					
Production materials are of appropriate durability and quality					
Print quality supports the readability and usability of the publication					

Figure 14-4: Section of judging form for evaluating promotional materials. *Reprinted courtesy of the Society for Technical Communication.*

What to Keep in Mind

- Use a dominant image—preferably one large graphic or text in a large-sized font.

- Include all the necessary information—the *who, what, when,* and *where*—and any contact information, if needed.

- Use a thick sans serif typefont—Impact, Ariel, or Comic Sans—one that will get attention and also reproduces well.

- Check for accuracy—since your design is limited to one page, any error (for example a misspelling or an incorrect date or time) will stand out and defeat the purpose of the flyer.

Figure 14-5: Flyer advertising STC-PMC's annual conference.

Brochures, Data Sheets, and Overviews

Purpose

Brochures, data sheets, and overviews contain marketing messages that inform potential clients about features and benefits of the group, products, or services. They also include a "call to action" by supplying a phone number or e-mail address to get additional information, arrange an appointment, or order the product itself. The design and layout, especially of brochures, need to be exceptionally attractive; and the colors used should add to the appeal and unify the overall design.

Content

Usually, brochure, data sheets, and overviews have sections where the main features of a product or the benefits to the customer are highlighted. These promotional pieces sometimes include a listing of product specifications on the back panel or at the end of the text. Because they contain information that is unlikely to change soon, they are often printed on expensive paper and contain attractive photos and graphics.

See Figure 14-6 for an example of a data sheet detailing APL Logistics' services for consumer packaged goods.

What to Keep in Mind

- Make sure your marketing message is clear and concise.
- Check that your design is unified and appropriate for your purpose.
- Be sure that you balance the features with the benefits to customers or clients.
- Check that captions and callouts are consistent and accurate.

Catalogs

Purpose

Catalogs are pre-sales publications that allow your potential customer to learn about and order your products. They can vary from being a collection of thin sheets with a few drawings printed in black ink only and held together with staples to being an elaborate book with photos and graphics printed in four colors on glossy paper with durable binding.

Content

In producing catalogs for consumer products, you'll probably use the latter type of catalogs to communicate more visually and show your products in an attractive manner. On the other hand, if you produce catalogs for industrial products, you'll probably use the former type and rely more on the products' specifications and features conveyed through words rather than though visuals.

What to Keep in Mind

- To make your catalog accessible to the average user, be sure that your index is accurate, complete, and easy to use.
- Mix graphics with text, even if the graphics are simple line drawings. To produce a catalog with all text is overwhelming— and a picture can sell your product a lot faster than a description using words alone.
- Title your catalog clearly so that the name actually reflects the contents.

Increasing Retail Sector Efficiency

Global Supply Chain Management

Supply chain management and logistics technologies are reshaping how consumer packaged goods (CPG) manufacturers respond to the needs of today's consumer-centric, demand-driven marketplace

The most successful CPG companies are building relationships with third-party logistics (3PL) providers such as APL Logistics to expand supply chain visibility and realize economies of scale in transportation and distribution.

Consumer Packaged Goods is a Core Competency

The CPG team at APL Logistics understands the competitive need to continuously reconfigure your supply chain to meet changing consumer purchasing patterns. Our team also helps drive down your cost of real estate, inventory, warehousing, transportation, and other non-revenue-producing supply chain assets.

We have designed and managed systems for a wide range of CPG manufacturers, including food, beverage, health and beauty aids, over-the-counter pharmaceuticals, household products, tobacco, and paper products. Our CPG team identifies opportunities in your supply chain that offer performance improvements through:

• Multi-country sourcing and order consolidation

• Global transportation management

• Deconsolidation alternatives

• Technology integration

APL Logistics also manages distribution of retail point of purchase (POP) displays and other promotional materials. For operations that require food-grade sanitation, our facilities consistently achieve superior marks from inspectors representing the American Institute of Baking (AIB).

Leading-edge Logistics Technology

In North America's big-box retail environment, operational and technological innovations are vital to revenue growth for both manufacturers and retailers.

APL Logistics provides unsurpassed control over order processing, inventory management, warehousing and transportation through industry-leading information technology. From the factory floors to the store shelves, APL Logistics puts vital information at the fingertips of every partner in your supply chain.

Designed for multi-platform compatibility, Our information technology solutions provide exceptional application quality. Implementation risk is reduced through proven interface compliance with SAP, Oracle, Baan, and other enterprise resource planning (ERP) systems. The result is a real-time picture of orders, inventory, and shipment data through a secure Web-based portal.

Win More Market Share

APL Logistics' mission is to forge a partnership that enables your company to refocus on its core competencies as we help bolster margins through improved operations efficiency and eliminate costly consequences of excess inventory or stock-outs. APL Logistics' value-added services are customized to serve the specialized need of the CPG industry.

Product Tracking — Through APL Logistics' WMS technology, product is monitored by expiration date, product and lot numbers, which reduces returns because inventory is rotated to avoid shipments of products nearing expiration. If a product recall is ever necessary, WMS can instantly display inventory status.

Figure 14-6: Data sheet on consumer packaged goods. *Reprinted courtesy of APL Logistics.*

Distribution, Manufacturing and Transportation Services

- Consolidation
- Cross-border services
- Custom packaging and labeling
- Customs brokerage services
- DC bypass
- Deconsolidation
- Dedicated contract carriage
- Delivery optimization
- Direct to store
- Distribution
- Facility design, sale, and lease-back
- Freight payment and audit
- Inventory management and financing
- JIT delivery, pickup, and replenishment
- Kitting
- Merge-in-transit
- Multi-modal Optimizations
- Packaging
- Quality assurance
- Real estate
- Retail displays (POP)
- Returns management
- Routing design and carrier negotiations
- Sequencing
- Sourcing/vendor management
- Sub-assembly
- Transportation Management
- Warehouse management
- Yard management

Custom Packaging — Products are configured through coupon insertion or special promotion stickers, bundling, and shrink-wrapping in APL Logistics distribution centers.

Retail Displays — POP displays are built so they only need to be unwrapped and placed by store personnel.

Delivery Optimization — Assures products are on retailers' shelves through delivery techniques such as multi-vendor consolidation, direct to store delivery, and just-in-time delivery, pickup, and replenishment.

Returns Management — Close the loop on your supply chain by processing returned goods and credit data. APL Logistics will inspect and sort for restock.

Serving Some of the Biggest Names in CPG

APL Logistics provides the supply chain expertise to engineer innovative, cost-effective solutions for some of the top names in consumer packaged goods.

Birds Eye Foods. As one of America's largest agricultural processing and marketing cooperatives, Birds Eye distributes a variety of branded, private label and food service products. Their most familiar labels include Comstock (fruits and pie fillings), Brooks (chili and related products), McKenzie's (frozen vegetables and fruits), and Pops Rite popcorn. APL Logistics provides value-added distribution and freight management services for a distribution center that serves four food processing plants within a 50-mile radius. During peak packing seasons, the center operates 24 hours/six days a week to receive, inventory, label, pack, stage, and ship an annual volume of more than 11 million cases.

Colgate-Palmolive. A global leader in oral-care products (mouthwashes, toothpaste and toothbrushes) Colgate sells more than 40 percent of the world's toothpaste. Colgate is also a major supplier of personal-care (deodorants, shampoos and soaps), household cleaning, and pet nutrition products. APL Logistics began its logistics partnership with Colgate in 1988 by managing its Atlanta distribution center. Today, the partnership has expanded to include operations in California, Georgia, Ohio, Texas, Mexico, and Puerto Rico, handling 600+ SKUs and more than 40 million cases of products annually. APL Logistics was an active partner in Colgate's successful SAP software implementation, which included an interface with APL Logistics' WMS software and the introduction of radio frequency technology into each distribution center.

The Kellogg Company. The world's largest maker of ready-to-eat cereals, the Kellogg Company boasts sales in excess of $10 billion. The company makes 12 of the top 15 cereals in the world and produces other grain-based convenience foods including cookies, crackers, toaster pastries, cereal bars, frozen waffles, and meat alternatives, manufacturing in 17 countries and marketed in more than 180 countries around the world. Kellogg's addition of the Keebler Foods line of business more than tripled the product picking volume for Kellogg's two California distribution centers. To successfully integrate Keebler Foods into the Kellogg facilities, APL Logistics reengineered the warehouse, including adding racks to facilitate pick line efficiency. A new warehouse management system with radio frequency capabilities was implemented to meet the increased order demand and still maintain inventory accuracy. Kellogg named APL Logistics "Top Gun" for 2005—the company's highest honor for retail warehouse performance among the third-party operators of its U.S. distribution centers, as the company maintained a 99.9+ percent inventory accuracy rate, 99 percent on-time shipping, and 98 percent order fill rate.

For more information, contact your APL Logistics Account Manager or visit www.apllogistics.com.

APLLMC-702 © APL Logistics apll_CPG.pdf 040407

Figure 14-6: Data sheet on consumer packaged goods. *Reprinted courtesy of APL Logistics. (continued)*

Checklist

	Do You Know	Yes	No
1.	the benefits of creating a style guide?		
2.	what areas are usually covered in style guides?		
3.	the benefits of creating a quick reference guide?		
4.	some of the criteria used to judge reference guides?		
5.	the characteristics of flyers?		
6.	why flyers are usually produced inexpensively?		
7.	what "call to action" messages are in brochures?		
8.	what distinguishes brochures from other types of promotional materials?		
9.	why catalogs are considered promotional materials?		
10.	which types of catalogs contain more visuals?		

Exercises

1. Discuss reasons for creating a style guide for an organization. What problems might arise if you don't have one?

2. Prepare an outline for a style guide. What categories would you include and exclude?

3. Find a quick reference guide, such as one for your telephone or printer; and write a brief evaluation of it as an effective document. Are the instructions clear? Are the visuals helpful? Include answers to these and similar questions in your evaluation.

4. Create a quick reference guide for a process that you're familiar with, using icons and symbols to help reduce the amount of text.

5. Find two or three flyers that are posted where you work or attend classes, and evaluate them. Do they attract attention? Do they contain all the needed information? How might they be improved?

6. Create a flyer for a future event with a dominant graphic that will attract viewers. Be sure to include all the needed information—date, time, place, and any costs.

7. Evaluate the overview below (Figure 14-7) to see that it conveys all the necessary information.

8. Evaluate the data sheet in Figure 14-6. Identify the features and benefits to the customer. Are they the same or different? Is the call to action clear? Can it be improved?

9. Create a brochure for an organization with which you are affiliated. Include the benefits and features, contact information, and several graphics.

10. Bring to class a catalog that you've received for books, clothes, electronic products, or some other items. Working in groups with other students, discuss what you think is attractive about it as a sales document and how you might improve it.

APL Logistics Overview
Global Supply Chain Management

Among the foremost providers of international integrated supply chain solutions, APL Logistics combines origin and destination logistics with multi-modal transportation services. Our offerings are enabled by industry-leading information management and data connectivity tools that provide increased product visibility and better inventory management to meet changing market conditions.

APL Logistics Services

- **Supply Chain Design** — Design of logistics networks that deliver on schedule, reduce cycle times, and reduce the administration cost of managing multiple vendors.
- **Consolidation** — Managing the movement of products to distribution centers and retail outlets worldwide, including the delivery of documents and electronic information critical to inventory management.
- **Vendor Management** — Local experts work directly with your vendors so your products are loaded, packaged, and delivered precisely how, when, and where your customers need them.
- **Manufacturing Support** — Integrating specialized facilities, operations, and information technology to provide manufacturing support services including kitting, sequencing, subassembly, just-in-time (JIT) delivery, and quality control services.
- **Global Freight Forwarding and Management** — Includes ocean freight, expedited freight, dedicated contract carriage, intermodal services, yard management, and customs brokerage.
- **Deconsolidation** — Includes expediting, delaying-in-transit, or transloading import shipments to reduce inland transportation costs and simplify your supply chain.
- **Warehousing and Distribution** — Improve inventory management, reduce operating costs, and speed order cycle times through contract or flexible space options. Other services include DC bypass, merge-in-transit, and vendor managed inventory.
- **Information Management Solutions** — Shipment visibility and event management tools including See Change Services and Shipment Tracking, RFID, warehouse management systems, and transportation and freight optimization.

Major Industries Served

- Automotive
- Electronics/Hi-Tech
- Consumer Durables
- Retail/Apparel
- Consumer Packaged Goods
- Heavy Equipment/Industrial

Figure 14-7: Overview detailing services to customers. *Reprinted courtesy of APL Logistics.*

CHAPTER FIFTEEN

Oral Presentations

Skills in oral communication are more valued than ever before. Recruiters at colleges and universities, in fact, value oral communication skills as much or more than writing and computer skills when considering future employees.[1] This chapter focuses on developing your oral communication skills by helping you plan, prepare, and deliver effective presentations, as well as integrate visuals within them. The chapter also looks at several forms that these presentations can take: speeches, PowerPoint® slide shows, podcasts, and multimedia presentations such as video blogs (vlogs) and screencasts.

Objectives

- ✓ Learn how to prepare and deliver effective oral presentations.
- ✓ Learn about the different forms for these presentations.
- ✓ Consider how to prepare or deliver each form.
- ✓ Consider how to integrate visuals and text effectively within them.

Preparation

In planning your presentation you need to analyze your audience and determine your purpose. You also need to consider your time limits and allow sufficient time for each of the major points you want to cover. You also want to integrate graphics and allow time to practice the presentation.

Analyze your Audience

Audience analysis is important for any form of communication, but with oral presentations it's even more important because your audience is physically present. Before beginning a speech, review the list of considerations for audience analysis given in Chapter One.

If the audience is not as knowledgeable as you are on the subject, reconsider your choice of words to ensure that all the terms are clear, and supply more details than you would for an audience who shares your background. If your analysis of the audience shows that they will probably be unreceptive to your presentation if you are direct about what you're advocating, you might reorganize it and consider a more indirect approach.

---

1   I conducted a survey in 2004, sending out questionnaires to 250 employers who recruit at La Salle University in Philadelphia, PA. Of the 59 respondents, the majority stressed the importance of oral communication and ranked it higher than written communication in answering a question about which skills among graduates did employers value the most.

Determine your Purpose

As explained in Chapter One, determining the purpose of any form of communication is linked closely to an analysis of your audience. With an oral presentation **ask yourself what you want to accomplish** by the end of your delivery. If your listeners are likely to be receptive, your purpose might be to inform or motivate. If they are hostile or indifferent, your purpose might be to have them reconsider their position or persuade them to adopt yours.

Consider your Time Limits

Knowing the amount of time you have is crucial for organizing and developing your points. If you're limited to 5-10 minutes rather than 20-25 minutes, you can cover only a few points and will be unable to go into extensive detail with any one of them. With a short presentation, be sure to allow at least one or two minutes for an introduction and conclusion so that you have enough time for these parts.

Outline your Major Points

Take the time to list the major points in the body of your speech, and **consider putting each on index cards.** If you're uncomfortable with cards, possibly for fear that you may drop them or put them in the wrong order, then **print your major points on a sheet of paper** that you can rest on a podium. **Consider writing a brief introduction before your points** because you are more likely to be a little nervous at the beginning of your speech. Also **consider writing a concluding sentence** after your last point so that you can end effectively.

Select Graphics

Select graphics carefully. Select those that are readily understood or recognized and clearly delineated. **Be sure that they are large enough to be seen** by all people in your audience, especially those in the back of the room. If they're complex or too small, consider duplicating and distributing them individually.

Integrate Graphics

Consider at what point in your presentation you want to refer to your graphics. If they're slides with complementary text, then **refer to them as you touch upon the points in the text they illustrate.** If they're separate, be sure to pause and allow enough time for the audience to examine them.

Practice, Practice, Practice

"Practice, practice, practice," the comic reply to the question about how to get to New York's Carnegie Hall, is good advice for anyone preparing a speech. When you've organized everything and considered how to develop your points, you need to practice aloud.

- **Practice the speech before a friend or acquaintance** who will not only substitute for your live audience but who also can offer critical feedback about how to improve your presentation.

- If no such person is available, **practice in front of a mirror or, if possible, use a recording device.**

- If you don't have these options or are uncomfortable using them, then **practice aloud alone**—preferably in a room similar in size to the one you'll be speaking in. Of course, if you can practice in the actual room itself before the presentation, you'll increase your sense of confidence and comfort.

Decide What to Wear

How you look will affect not only how your audience views you but also how you view yourself. Choose clothing that you feel successful wearing. Ideally, **your clothing should be both comfortable and attractive**; however, more importantly, it should not divert your audience from focusing on what you're saying.

Delivery

Once you've prepared your presentation fully, you'll be more confident as you prepare to deliver it. Before you speak, however, keep in mind the following:

Consider your Posture and Expression

As you prepare to speak, check your posture and appearance. You want to **stand up straight but not appear rigid**. Try to stand still and not move around too much as you speak. Walking back and forth can be tiresome to watch. Using hand gestures, however, can actually be quite effective—if not overdone. Certainly, use them if they work to create a sense of inclusiveness with your audience and help you establish a rapport with them. Also check your expression. You want to look confident, so relax and smile to give the impression that you're actually enjoying your own presentation.

Look at your Audience

Look at your audience not only at the beginning of your presentation but throughout all of it. **In a room of 20-25 people, you want to make eye contact with everyone at least once.** If more people are in the room, try to make eye contact with as many as you can; and try to give the impression that you're looking at everyone individually.

Refer to your Visuals

Whatever visuals you have, try to integrate them into your delivery. **As you refer to visuals, point to them, hold them up, or interact in some other way with them.** Because these visuals are part of your presentation, like the people in your audience, you too need to be consciously interacting with them.

Modulate your Voice

Certainly, you want to **speak loud enough for those in the back of the room to hear you**, but you also want to modulate your voice so that it's soft as well as loud at times. Additionally, you want to consider your speed. You might speak fast in going over some minor details, but then slow down to emphasize a major point. If your presentation is given at the same speed and volume throughout, it'll seem unnatural and boring.

Hold Questions to the End

Although you want to encourage questions, you want to hold them to the end. While taking questions during the presentation can give the impression of having a dialogue with the audience, some people, those not asking questions, might consider them as disruptive, especially if these questions are unrelated to their own concerns. Also, **taking questions during the presentation can sometimes make it difficult to maintain the flow of what you're saying** and can also make it difficult once you stop to return to the place where you left off.

Conclude Strongly

When finishing, you want to end on a positive note. Slow down, speak firmly, and give your final statement in such a way that your audience knows your speech is concluded. **If you haven't already told your audience that you'll take questions at the end, do so now.**

Ask for questions and then wait politely, even if no one asks a question immediately. When you get one, try to answer it directly. If you're unclear about what is being asked, ask the questioner to repeat the question; and if the questioner speaks too softly when asking a question, don't hesitate to repeat the question more loudly so others can hear it. If you don't know the answer to a question, admit you don't know it and offer to get back to the questioner later with the answer, if doing so is possible. Keep track of your time. When your allotted time is nearly exhausted, announce that you can take only one more question or that you can elaborate more only on the current question. And then **conclude by thanking your audience for their time and attention.**

Using Oral Presentation Skills

Speeches

Speeches are probably the most traditional form of oral presentations. While most likely you'll give speeches at different times and on different occasions throughout your career, most of them can be classified as one of three types.

Impromptu

Impromptu presentations are usually short and allow you to speak about something you're knowledgeable about with little or no preparation. When asked to give an impromptu report on an item of business, try to prepare yourself mentally. Think about what you want to achieve (your purpose) and a few points that will help you achieve this goal. If asked to give an impromptu presentation, in most cases you'll be knowledgeable not only about the subject but also about your audience. However, take a few moments to look around at your audience, and consider whether they share your familiarity with the subject and if they're receptive to what you're saying.

Scripted

Opposite to the impromptu speech is one that is carefully written and then read or memorized. While this type of presentation allows for thorough preparation ahead of time, it can be deadly if delivered in a stilted, unnatural manner. If you memorize a script, deliver it almost as if you're speaking impromptu. However, keep in mind that memorizing a long script can be extremely challenging—and if you get confused and forget where you are in this script, the result can be disastrous. If you read from your script—which is preferable to memorizing it—continually look at your audience and use body language to convey as natural a manner of speaking as possible.

Extemporaneous

The most popular type of speech is the one you prepare with notes or outline beforehand, but that you write out only partially. Certainly, you can write out word-for-word your introductory or concluding remarks or any quotations or factual information you want to include. However, an extemporaneous speech is quite effective when you present it in a natural manner. Also, because you have sufficient time to prepare it, it's usually more organized, more detailed, and consequently more effective than an impromptu one.

Poster Sessions

Poster sessions provide occasions for you to use your oral presentation skills, not as you would in giving a speech but as you would in answering questions at the end of it. According to a Colorado State University Web site, "a Poster Session allows viewers to study and restudy your information and discuss it with you one on one."[2] While you obviously need to put more time into preparing your visuals than in what you say, you nevertheless want to be able to discuss your research in an organized manner and thus should spend some time preparing answers to questions that might arise.

Types

Most people think of posters as those displaying research on printed pages attached to *cardboard or foam panels* standing upright on tables (see Figure 15-1) or consisting of large sheets of paper that are taped or pinned to walls. However, while these traditional posters are still popular, more likely you'll attend poster sessions that include electronic presentations, those that display information on *laptop computers* using Internet pages, PowerPoint® slides, or screens from other software (see Figure 15-2). The benefit of these displays is the amount of information that you can present in a small space. Using links, you let the viewers go to different sites for more information and choose what they want to read about. Keep in mind, though, that just as you would stand beside your cardboard poster, you'll want to stand beside your laptop to show your electronic poster and answer questions from viewers (see Figure 15-3).

Content

Colorado State University's Website gives an example of information to include on a poster under the following headings:

> Statement of the Problem
> Methods for Collecting Information
> Results
> Key Findings[3]

2   "Definition of a Poster Session" Presenting Poster Sessions. Colorado State University. 31May 2007.<http:// writing. colostate.edu/guides/speaking/poster/pop2a.cfm>.

3   "Example Poster." Presenting Poster Sessions. Colorado State University. 21 July 2006.<http://writing. colostate.edu/guides/speaking/poster/list17.cfm>.

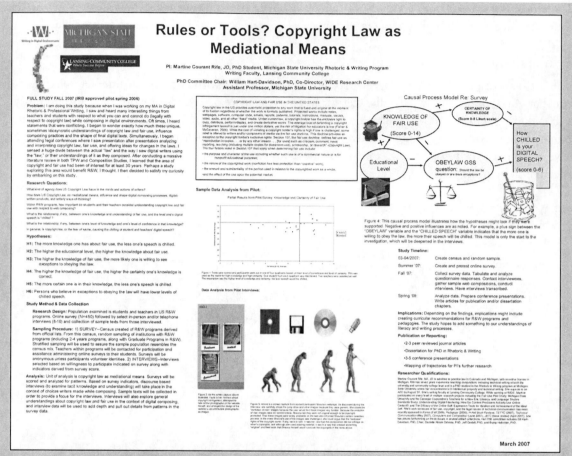

Figure 15-1: Cardboard poster displayed at an Association of Teachers of Technical Writing Conference in New York.

Figure 15-2: Electronic poster showing the development of a game presented during a poster session for undergraduate research projects at La Salle University, Philadelphia.

Figure 15-3: Student researcher stands by her laptop ready to discuss her poster created with Microsoft® PowerPoint® slides.

Whatever organization you use, however, your information should answer the common journalistic questions: what, when, where, why, and how. In a Web site entitled "How to Present a Poster Session," the author explains that the answers to these questions "are often listed as INTRODUCTION & RATIONALE, RELATED LITERATURE, METHOD, FINDINGS, IMPLICATIONS."[4]

The following are tips to help you prepare a poster.

- **Do** focus on the main idea of the study, and use text and graphics to make this idea prominent.
- **Do** provide a supplemental handout with material that is not given on the poster itself.
- **Do** include links on an Internet poster so that the viewer can click onto them for more information.
- **Do** supply contact information on the poster or accompanying handouts.
- **Don't** have too much text.
- **Don't** use graphics that are too small to see.
- **Don't** use graphics that are unclear or difficult to interpret.
- **Don't** use colors that are too similar or that clash—and thus distract from the content.

---

4   "How to Present a Poster Session." 31May 2007. <http://educ.queensu.ca/~ar/poster.htm>.

PowerPoint® Slide Shows

Most likely you've given a PowerPoint® presentation or someday will be giving one. So popular is this presentation software that Microsoft claims that there are 30 million or more such presentations given daily around the world.[5] Despite its popularity, many of these presentations are poorly constructed and delivered. They are often the result from poor planning and neglect of the basics that go into creating any form of communication. To avoid putting your audience to sleep or just boring them with mind-numbing slides, follow the advice given earlier under "Preparation" that involves determining purpose and analyzing audience. Also, consider the following six tips:

Outline and create a storyboard

Rather than putting an outline of your major points on note cards, **consider creating an outline on a page divided into two parts: one for text and one for graphics.** Then on the side for text, list major headers that you can translate into titles for individual slides. Divide these headers into sub-headers, but keep in mind that you need at least two sub-headers. If you have only one sub-header, then go back and reword the header so that its meaning encompasses what you originally planned to put under it.

Since your presentation is a visual one, consider creating a storyboard by sketching a graphic next to each of the major headers in your outline. Use a separate page for each slide, placing the text on one side and the sketch or name of the graphic on the other. In your "story," be sure to include an introduction, conclusion, and a few transitional slides.

Limit the amount of text, and choose appropriate graphics

Limit the amount of text on a slide to 25–35 words, divided into no more than six lines, if possible. Too many words on a slide reduce the type size and make reading difficult. Also, you want to use fewer words so that you can elaborate verbally on what you've written. In selecting graphics, consider the tone you want to convey: for a more formal tone, you might use only photos; whereas for an informal one, you might use cartoons.

Use a sans serif font

Many presenters just convert text from pages typed in a default font like Times Roman or Cambria to the text on their PowerPoint® slides. While this serif font is quite readable on the printed page, on a screen it's more difficult to read than a sans serif font like Arial, Calibri, or Impact. Select a font that has strong, delineated lines so that the words stand out and everyone in the room can read them.

---

5   Goldstein, M. "It's Alive! The Audience, That is, But Some Presenters Don't Seem to Know It." *Successful Meetings* 52.2 (2003): 20 referenced in Debbie D. DuFrene and Carol M. Lehman. "Concept, Content, Construction, and Contingencies: Getting the Horse Before the PowerPoint Cart." *BCQ* 67.1(March 2004): 84.

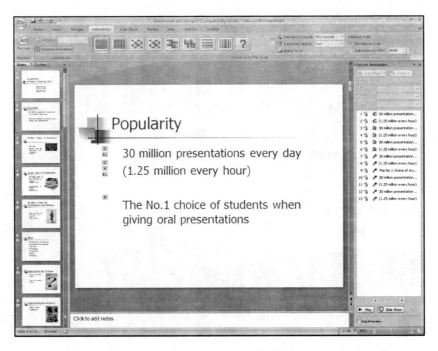

Figure 15-4: PowerPoint® slide with options for custom animation.

Use colors with sufficient contrast

While you might want to change the color of your type, keep in mind that light colors like yellow or gold that look fine on your small computer screen can seem washed out and extremely difficult to read when you project them on a larger room screen. **If you do use a light-colored type, consider highlighting it with a "shadow"** to have more contrast.

Use animation and transitions sparingly

PowerPoint® software offers a variety of animated schemes and transitional effects that can enliven your presentation (see Figure 15-4). Animation schemes range from subtle ones, such as "fade" or "dissolve," to bolder ones, such as "pinwheel" and "boomerang." While these animated effects and transitional ones, such as vertical or horizontal blinds, can add a nice touch while introducing your slides, **they can be distracting and annoying if overused**. Keep in mind that if you're going to use an animated scheme or a transition, you want to choose one that fits well with your subject and the tone of your presentation. Preferably, use one such scheme or transition consistently throughout, or use one scheme and then add another one for variation. Avoid using several different schemes and randomly switching from one to another.

Having your words fly in from the left on a slide, or having your slides change from one to the other like vertical blinds can work well. But having your words fly in from the bottom and then from the left and right or having screens change as blinds and then dissolve and fade will only distract your audience.

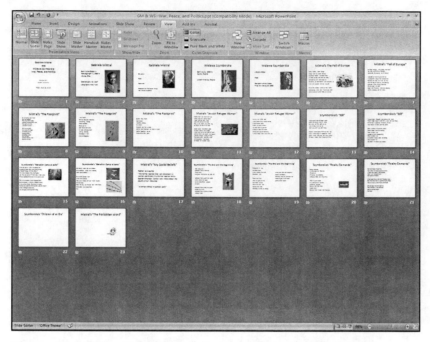

Figure 15-5: Slide Sorter view of a PowerPoint® presentation.

If you create effects, check to see before giving the presentation that they are working properly by clicking on the star icon in the lower left corner of each slide in "Slide Sorter" under "View" (see Figure 15-5). If any transition fails to function properly, make the necessary changes.

Edit for accuracy, consistency, clarity, and conciseness.

In editing, review the entire presentation first for accuracy in content and then for organization. **Check first that all your facts are correct.** Then check the overall arrangement of the slides in the "Slide Sorter" view. If any slide seems out of place, cut and paste it where you want it to appear.

After reviewing the overall presentation, check individual slides for accuracy and consistency with punctuation, capitalization, color, type font, and effects. Check also for clarity and conciseness. Make sure that what you want to say will be clear to your viewer. To improve clarity, check if you can eliminate any unnecessary words, condense phrases, or delete extra graphics.

Podcasts

Podcasts are another form of oral presentations that you can record in audio files and upload to the Internet, so that they are available for downloading to MP3 players. In doing so, you can use your oral presentation skills to reach large audiences at different times and places who, in turn, can listen to you at their convenience. Tom Johnson, founder of Tech Writer Voices (www.techwritervoices.com), a podcast for technical writers, points out that podcasting allows you to provide content from live events that have a limited number of participants present. "In contrast with webinars, teleseminars, and conferences, podcasts remain permanently online," he explains. Johnson writes that unlike e-mail where messages compete with hundreds of others, "podcasting provides a unique one-on-one connection with your listener, allowing him or her to absorb the content more fully."[6]

Since it's a one-to-one connection, he argues that you don't need to be a professional broadcaster; but nonetheless, you do want to prepare what, in many cases, is a scripted presentation. To do so, consider the first four points discussed earlier under "Preparation" dealing with audience analysis, determination of purpose, setting time limits, and outlining major points. In addition, think about your tone or volume, and remember to modulate your voice.

Videoblogs

Since so many people are visually oriented, you might decide to videotape what you would normally record in a podcast or write in a blog. Videoblogging or "vlogging" consists of creating a script based on a specific topic which you deliver and record either with a digital camera or camcorder. Keep in mind that unlike the podcast, your appearance—not just your voice—is important. And while you want to follow the guidelines given earlier for preparation, you also need to follow those for an effective delivery.

Screencasts

Stephanie Cottrell Bryant, a Silicon Valley technical communicator, explains that in creating a videoblog, you might use PowerPoint slides as "visual breaks between a filmed 'talking head' and more dynamic presentation clips."[7] You can also add screenshots using screencasting software such as TechSmith's® Camtasia® or Adobe® Capture. Raymond K. Archee, who writes a column called "Computer -Mediated Communication" for the Society for Technical Communication's *Intercom*, hypothesizes that screencasting will become the future of technical communication and envisions user manuals being replaced by a combination of screenshots and video showing an experienced client demonstrating how to use a product or service. He points out that the name *screencasting* is not accurate, "since screen captures comprise only a small part of the kind of media that may be assembled to form an online video/animation."[8]

6   Johnson, T. "Podcasting: A New Layer of Communication." *Intercom* (January 2007): 13-16. 18 April 2008. <http://www.stc.org/intercom/PDFs/2007/200701_12-16.pdf.>.

7   Bryant, S. C. "Winning your Audience with Videoblogs"*Intercom* (September/October 2007): 25-27. 18 April 2008. <http://www.stc.org/intercom/PDFs/2007/20070910_25-27.pdf.>.

8   Archee, R.K. "Screencasting—the Future of Technical Communication?" *Intercom* (March 2008):39. 18 April 2008. <http://www.stc.org/intercom/PDFs/2008/200803_39.pdf.>.

Checklist

	Do You Know	Yes	No
1.	the difference between impromptu, scripted, and extemporaneous speeches?		
2.	why in preparing your presentation, you need to first consider your audience and purpose?		
3.	why it's important to consider time limits?		
4.	why you want to outline major points and write out introductory and concluding statements?		
5.	why it's important to consider complexity and size when selecting graphics?		
6.	what you need to consider regarding dress, posture, and expression when delivering your presentation in person?		
7.	what you need to consider regarding eye contact and using your voice?		
8.	why it's important to hold questions to the end?		
9.	what's important in answering questions and concluding?		
10.	what's important in preparing posters?		
11.	why creating a storyboard for a PowerPoint® presentation can result in an effective mix of text and graphics?		
12.	what colors and fonts to use for text on slides?		
13.	how many words and lines you should limit your text to on one slide?		
14.	what you should keep in mind when delivering a scripted podcast?		
15.	why your "personality" and appearance are important in a videoblog?		
16.	in addition to screen shots, what screencasting consists of?		

Exercises

1. Discuss which type of speech would be most appropriate for the following scenarios—impromptu, extemporaneous, or scripted:
   a. speaking after your name was unexpectedly announced as the winner of an award.
   b. speaking to a group of employees the day after you learned that you would need to motivate them either to increase their sales this month or face some layoffs.
   c. speaking at your former high school as the guest speaker at graduation.
   d. speaking on behalf of a friend who's seeking membership in an organization to which you belong.

2. Explain how you would handle the following situations when making your speech or Powerpoint® presentation:
   a. You enter the room and see a sparse audience scattered throughout the room—with many sitting in the last row and the first three rows vacant.
   b. During your presentation, someone raises her hand with a question; you stop and answer it, but the person continues to comment and ask additional questions.
   c. During your presentation, you notice the audience seems to nod in agreement and be involved in what you're saying, but people in the last two rows look puzzled and restless.
   d. You have handouts to accompany your presentation and give them out before beginning. While you speak, you notice people turning pages and reading the handouts and not fully listening to what you're saying.

3. Give a 3-5 minute impromptu speech on one of the following topics:
   a. why you chose your major
   b. what you like or don't like about your part-time job
   c. your favorite movie, TV show, book, sport, or place to relax

4. Prepare a 5-10 minute extemporaneous speech on a topic you're researching for one of your classes.

5. Critique the following PowerPoint® slide (Figures 15-6) from a presentation on paper-making. Consider the layout and design, use of type, and spacing as well as the writing itself. How would you improve it?

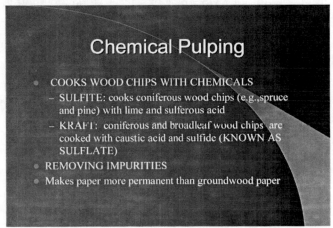

Figure 15-6: Slide on the pulping stage of paper-making.

6. Based on what you learned from the previous exercise, write a paragraph evaluating another slide on the paper-making process (Figure 15-7).

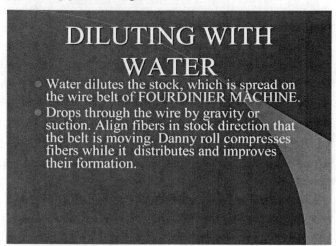

Figure 15-7: Slide on the diluting stage of paper-making.

7. Working with another student, list several advantages and disadvantages for creating both a traditional cardboard poster and an electronic one.

8. Using research for a project for one of your courses, create a poster to display it. What kind of poster would be best for you? What graphics would you include? What would be your focus?

9. Prepare a PowerPoint® presentation on the research you have completed for a project for this course. Have at least seven slides that incorporate both text and graphics. Present it to your classmates.

10. Write a paper (3-5 pages) evaluating a PowerPoint® presentation that another student gave in one of your classes. Did the student follow the guidelines outlined in this chapter? What do you think was effective about his or her presentation? How might it be improved?

11. Prepare a 5-minute script for a podcast on a subject with which you're familiar, and if the equipment and software is available, record it. In doing so, first explain what steps you took to prepare it, and what you tried to keep in mind when delivering it.

12. Prepare a 7-10 minute script for a videoblog or screencast on any subject. Include at least three PowerPoint® slides.

13. Go to a site that explains how to make videoblogs (www.Freelog.org), and write a brief summary of the main steps. Do you think the process is one that is easy or difficult to follow?

14. Look at Adobe's Creative Suite tutorials (www.CreativeSuitePodcast.com) which Adobe calls "videopodcasts." What forms of technology appear within them, and how effective are they in combining the various forms?

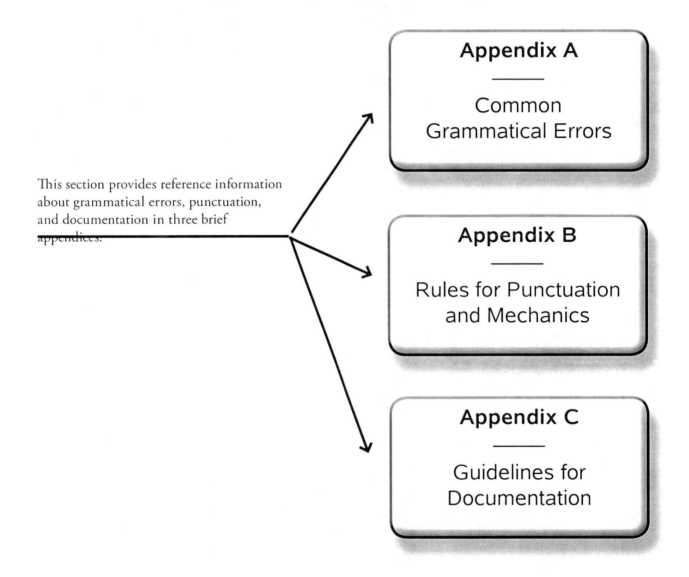

PART IV: Appendices

This section provides reference information about grammatical errors, punctuation, and documentation in three brief appendices.

Appendix A

———

Common Grammatical Errors

Appendix B

———

Rules for Punctuation and Mechanics

Appendix C

———

Guidelines for Documentation

APPENDIX A

Common Grammatical Errors

Fragments (Frag)

Fragments are incomplete sentences—phrases or clauses punctuated as complete thoughts. Though acceptable in speech and informal writing, they are unacceptable in Standard English with a few exceptions, one being lists.

To correct them, add the missing words or eliminate the conjunction that makes the clause a dependent one:

> **Fragment:** Because résumés are sent electronically.
>
> **Corrent:** Résumés are sent electronically.

Run-ons and comma splices (Run-on & CS)

A run-on sentence is really two sentences that are fused together without the punctuation needed to separate them (a period, semicolon, or comma with a coordinating conjunction):

> **Run-on:** Long reports are often spiral-bound they are then delivered to their readers.
>
> **Correct:** Long reports are often spiral-bound. They are then delivered to their readers.

A **comma splice** is a run-on sentence that is separated into at least two main clauses and punctuated with a comma rather than a period, semicolon, or coordinating conjunction:

> **Comma Splice:** Long reports are often spiral-bound, they are then delivered to their readers.
>
> **Correct:** Long reports are often spiral-bound; they are then delivered to their readers.

Subject-Verb Agreement (S-V Agr)

The subject and verb of a sentence do not agree if one is singular and the other is plural:

> **S-V Agr:** The <u>writer</u> along with her colleagues <u>have</u> reviewed the drafts.
>
> **Correct:** The <u>writer</u> along with her colleagues <u>has</u> reviewed the drafts.

Pronoun Errors

Reference-Agreement (Pron-Ref Agr)

A pronoun and the antecedent to which it refers do not agree if one is singular, and the other is plural; or if one is one case (for example, *possessive*), and the other is in another case (for example, *subjective*):

> **Pron-Ref Agr:** The <u>manager</u> accepted the <u>report</u>, and she forwarded <u>them</u> onto <u>their</u> boss.
>
> **Correct:** The <u>manager</u> accepted the <u>report</u>, and she forwarded <u>it</u> onto <u>her</u> boss.

Vague or Broad Reference (Vague Ref/Broad Ref)

A vague or broad reference error occurs when a pronoun does not refer back to one noun or pronoun as its antecedent, but rather refers back to the entire idea in the previous clause or to an antecedent that is only implied:

Vague/Broad Ref: We decided to release the new product on the first of the month, <u>which</u> pleased the manager.

Correct: We decided to release the new product on the first of the month. <u>This decision</u> pleased the manager.

Ambiguous Reference (Amb Ref)

A pronoun is ambiguous when it can refer to more than one antecedent:

Amb Ref: The <u>writers</u> gave the <u>editors</u> the drafts so <u>they</u> could meet their deadline.

Who needs to meet their deadline—the writers or the editors?

Correct: The writers gave the <u>editors</u> the drafts so the editors could meet their deadline.

The <u>writers</u> gave the editors the drafts so the <u>writers</u> could meet their deadline.

Dangling and Misplaced Modifiers

Modifiers need to be linked to specific nouns in sentences and placed beside the nouns that they modify. If the modifier just "dangles," supply a noun that it can modify, or rewrite the sentence. If you place the modifier too far away from the noun it modifies, rewrite the sentence to move it closer.

Dangling Modifier: <u>Pushing</u> firmly down on the brake pedal, the car will stop. (*who's pushing?*)

Correct: <u>Pushing</u> firmly down on the brake pedal, you will stop the car.

<u>Pushing</u> firmly down on the brake pedal stops the car.

Misplaced Modifier: He stopped the car in time to avoid an accident, having pushed on the brake pedal firmly with another vehicle.

Correct: Having pushed on the brake pedal firmly, he stopped the car in time to avoid an accident with another vehicle.

APPENDIX B

Rules for Punctuation and Mechanics

Commas (,)

Commas are used to separate elements in sentences, particularly in four instances:

To separate two independent clauses joined by a coordinating conjunction:

The manual was ready to be printed, but the product was not ready for delivery.

To separate an introductory dependent clause or long phrase from the independent clause that follows:

Although the manual was ready to be printed, the product was not ready for delivery.

To separate items in a series:

Check for mistakes with commas, apostrophes, and capitalization.

Separate parenthetical elements (elements that you can put in parentheses and remove from a sentence):

Mistakes, which can occur if you don't proofread your draft, can be expensive to correct after the manual is printed.

Avoid putting extra commas in your sentences:

Extra Comma: The proposal contained several items such as, features, benefits, costs, and references.

Correct: The proposal contained several items such as features, benefits, costs, and references.

Apostrophes (')

Possession

An apostrophe is mainly used to show possession. For singular nouns, add the apostrophe and an "s": the writer's draft.

For nouns that form their plurals with "s" or "es," add just the apostrophe: the writers' drafts.

However, for irregular plurals—nouns that form their plurals irregularly, you add the apostrophe and "s" to the plural: children's books or sheep's wool.

Contractions

An apostrophe is also used with contractions to indicate the omission of letters or numbers: for example, don't, haven't, I'll, I'm, or '76.

Plurals of letters and numbers

An apostrophe is also used to indicate the plural of lowercase letters and numbers as well as words in quotations and some abbreviations, for example those with two or more periods:

> Dot your i's, and cross your t's.
> We placed 6's and 7's at the bottom of the pages in the manuals where they were missing.
> How may "no's" did you count?
> Some of the science articles were written by M.D.'s.

Misplaced or Unnecessary Apostrophes

An apostrophe is misplaced when it's put after the "s" when it needs to be placed before it. Sometimes it needs to be omitted entirely. This problem occurs occasionally with the confusion between the contraction **it's** (it is) and the possessive adjective **its**:

> **Misplaced Apos:** We found the letter's envelopes.
> **Correct:** We found the letters' envelopes.
> **Correct:** What is its address?

Colons (:)

A colon is used mainly to introduce a series or to connect two independent clauses when the second one relates to or explains the first:

> The following items should be packed with the monitor: wires, power surge, stand, and instruction booklet.
> Checking the contents of the package is important: doing so ensures that the parts have all been sent.

Semicolons (;)

A semicolon connects two independent clauses. Often they are related in content, but they need not be: Install the red wire; then install the black one.

Capitalization (Cap)

Capitals are used for all proper nouns and the initial letters of the principal words of titles. Don't capitalize excessively. The down-style (using all lower-case letters) has been popular in recent years, and the trend is to capitalize minimally.

Abbreviations and Acronyms (Abbr)

Abbreviations are letters that represent words. Use only standard abbreviations, and be consistent with capitalizing or not capitalizing them. If you're unsure whether the abbreviation is standard, check your dictionary; and check whether it's capitalized or not. Some standard abbreviations such as those for morning or night (am/pm, a.m. /p.m., A.M. /P.M.) vary. Whatever form you choose, be consistent; and use it throughout your document.

Acronyms are made up of letters that each represents a word: SEPTA =Southeastern Pennsylvania Transportation Authority or OSHA=Occupational Safety and Health Administration. Be wary of overusing acronyms or using unfamiliar ones.

Numbers (Num)

In technical and professional communication, use words for numbers one to ten and Arabic numerals for all numbers over ten with a few exceptions:

- If you have two numbers in a sentence and one would normally be a word (a number under ten) and the other would be an Arabic numeral (a number over ten), use Arabic numerals for both: for example, 8 nuts and 13 bolts.

- If you're using a number to signify a quantity of items and the item is a number, write one number as a word and the other one as an Arabic numeral: for example, 2 one-inch screws.

APPENDIX B

Rules for Punctuation and Mechanics

Commas (,)

Commas are used to separate elements in sentences, particularly in four instances:

To separate two independent clauses joined by a coordinating conjunction:

The manual was ready to be printed, but the product was not ready for delivery.

To separate an introductory dependent clause or long phrase from the independent clause that follows:

Although the manual was ready to be printed, the product was not ready for delivery.

To separate items in a series:

Check for mistakes with commas, apostrophes, and capitalization.

Separate parenthetical elements (elements that you can put in parentheses and remove from a sentence):

Mistakes, which can occur if you don't proofread your draft, can be expensive to correct after the manual is printed.

Avoid putting extra commas in your sentences:

Extra Comma: The proposal contained several items such as, features, benefits, costs, and references.

Correct: The proposal contained several items such as features, benefits, costs, and references.

Apostrophes (')

Possession

An apostrophe is mainly used to show possession. For singular nouns, add the apostrophe and an "s": the writer's draft.

For nouns that form their plurals with "s" or "es," add just the apostrophe: the writers' drafts.

However, for irregular plurals—nouns that form their plurals irregularly, you add the apostrophe and "s" to the plural: children's books or sheep's wool.

Contractions

An apostrophe is also used with contractions to indicate the omission of letters or numbers: for example, don't, haven't, I'll, I'm, or '76.

Plurals of letters and numbers

An apostrophe is also used to indicate the plural of lowercase letters and numbers as well as words in quotations and some abbreviations, for example those with two or more periods:

Dot your i's, and cross your t's.

We placed 6's and 7's at the bottom of the pages in the manuals where they were missing.

How may "no's" did you count?

Some of the science articles were written by M.D.'s.

Misplaced or Unnecessary Apostrophes

An apostrophe is misplaced when it's put after the "s" when it needs to be placed before it. Sometimes it needs to be omitted entirely. This problem occurs occasionally with the confusion between the contraction **it's** (it is) and the possessive adjective **its**:

Misplaced Apos: We found the letter's envelopes.

Correct: We found the letters' envelopes.

Correct: What is _its_ address?

Colons (:)

A colon is used mainly to introduce a series or to connect two independent clauses when the second one relates to or explains the first:

The following items should be packed with the monitor: wires, power surge, stand, and instruction booklet.

Checking the contents of the package is important: doing so ensures that the parts have all been sent.

Semicolons (;)

A semicolon connects two independent clauses. Often they are related in content, but they need not be: Install the red wire; then install the black one.

Capitalization (Cap)

Capitals are used for all proper nouns and the initial letters of the principal words of titles. Don't capitalize excessively. The down-style (using all lower-case letters) has been popular in recent years, and the trend is to capitalize minimally.

APPENDIX C

Guidelines for Documentation

In the corporate world, you'll seldom include a list of secondary sources when producing technical publications. However as a student, your instructor might require that you submit a list of them with your technical communication assignments. Most likely your instructor will require that you follow the style guidelines established by the Modern Language Association (MLA) or the American Psychological Association (APA). Samples of both are shown here.

MLA

In-text Citation

For sources that you paraphrase or quote within the text, include in parentheses the author (if known) or the name of the source (if the author is unknown) and the page number:

> "In many colleges and universities, only three courses in the undergraduate curriculum introduce the relationship between ethics and communication—business communication, technical communication, and interpersonal communication" (Barnes and Keleher 144).

If you give the author's name in the sentence, you don't need to repeat it in the parentheses:

> Barnes and Keleher explain that "only three courses in the undergraduate curriculum introduce the relationship between ethics and communication—business communication, technical communication, and interpersonal communication" (144).

Works Cited

At the end of a document containing sources, list all the works that have been cited. The following are sample entries for some of the most common **print** sources:

Book

Tufte, Edward R. *The Visual Display of Quantitative Information.* 2nd ed. Cheshire, Conn: Graphics Press, 2001.

Journal

Barnes, Michael C., and Michael Keleher. "Ethics in Conflict: Making the Case for Critical Pedagogy": *Business Communication Quarterly* 69.2 (2008): 144-57.

Magazine

Roberts, Johnnie L. "Watching the Watchers" *Newsweek* July 17, 2007: 38-39.

Newspaper

Fishman, Charles. "Corporate Secrecy Creates A Skewed Perspective." *The Philadelphia Inquirer.* July 9, 2007: D1 & D5.

The following are entries for some common **electronic** sources:

Internet site

"Competition Guidelines." *Society for Technical Communication.* 19 July 2007 <http: www. stc.org>.

Newspaper article found in database

"Technology Briefing." *The Washington Post* 31 January 2006: D.05. ProQuest. La Salle University Connelly Library, Philadelphia, PA. 26 December 2007.

Journal article found in database

Hung, Chen-Ming, and Lee Feng Chien. "Web-Based Test Classification in the Absence of Manually Labeled Training Documents." *Journal of the American Society for Information Science and Technology* 58.1 (2007): 88. ProQuest. La Salle University Connelly Library, Philadelphia, PA. 26 December 2007.

APA

In-text Citation

For sources within the text of a document, give the author's last name and the year of publication within parentheses at the end of a sentence. If the text is a quotation, write p. (or pp.) and the page number(s):

"In many colleges and universities, only three courses in the undergraduate curriculum introduce the relationship between ethics and communication—business communication, technical communication, and interpersonal communication" (Barnes & Keleher, 2008, p. 144).

If you give the author or authors' names within the sentence, then add the year of publication in parentheses after the name or names. If the text is a quotation, put "p" and the page number in parentheses at the end of the quotation:

Barnes and Keleher (2008) explain that "only three courses in the undergraduate curriculum introduce the relationship between ethics and communication—business communication, technical communication, and interpersonal communication" (p. 144).

References

The following are entries for some common types of **print** sources:

Book

Tufte, E. R. (2001). *The visual display of quantitative information* (2nd ed.). Cheshire, CT: Graphics Press.

Journal

Barnes, M. C., & Keleher, M. (2008). Ethics in conflict: Making the case for critical pedagogy. *Business Communication Quarterly, 69(2),* 144-157.

Magazine

Roberts, J. L. (2007, July 17). Watching the watchers. *Newsweek,* 38-39.

Newspaper

Fishman, C. (2007, July 9). Corporate secrecy creates a skewed perspective. *The Philadelphia Inquirer,* pp.1-5.

The following are entries for some electronic sources:

Internet site (no author identified)

Society for Technical Communication (2007). *Competition guidelines.* Retrieved February 5, 2007, from STC Website: http:// www.stc.org.

Journal article found in database

Hung, C., & Chien, L. (2007). Web-based text classification in the absence of manually labeled training documents. *Journal of the American Society for Information Science and Technology,* 58(1), 88. Retrieved December 26, 2007, from ProQuest database.

INDEX